Emerging Pattern in Plant Genomics and Marker Assisted Breeding

Lavudya Sampath, from Warangal, Telangana, holds a B.Sc. in Agriculture from UAS, Raichur, Karnataka, and M.Sc. in Genetics and Plant Breeding from AAU, Anand, Gujarat. He is pursuing a Ph.D. in Plant Breeding and Genetics at Tamil Nadu Agricultural University, Coimbatore, with an ICAR-SRF scholarship and CSIR-UGC NET qualification in life sciences. Sampath has authored academic publications and participates in several conferences.

Tushar Arun Mohanty, from Odisha, holds a bachelor's in agriculture from Siksha "O" Anusandhan University and a master's in Genetics and Plant Breeding from RPCAU, Pusa, Bihar. He worked as an Assistant Professor at MITS Institute of Professional Studies in Rayagada, Odisha, and is currently pursuing a Ph.D. at TNAU, Coimbatore, actively engaged in publications and conferences.

Santhosh V., from Telangan, holds a B.Sc. in Agriculture from PJTSAU, Hyderabad and M.Sc. in Genetics and Plant Breeding from AAU, Anand, Gujarat. He was previously worked as SRF in ICAR-IIOR, Hydearabad, Telangana and is currently pursuing a Ph.D. at BHU, Varanasi, actively engaged in publications and conferences.

Swapnil Baraskar, from Ahmednagar, Maharashtra, holds a B.Sc. in Horticulture from DBSKKV, Dapoli, Maharashtra, and M.Sc. in Genetics and Plant Breeding from AAU, Anand, Gujarat. He is pursuing a Ph.D. in Plant Breeding and Genetics at PJTSAU, Hyderabad, with an ICAR-SRF scholarship, ICAR-NET and UGC NET qualification in environmental sciences.

Prajapati Maulik R., a native of Mahesana, Gujarat, holds a B.Sc. (Hons) Agriculture and M.Sc. in Genetics and Plant Breeding from NAU, Navsari, Gujarat. He is currently working on Salinity tolerances in Rice as part of his Ph.D. programme at the Regional Rice Research Station in Vyara, Department Genetics and Plant Breeding NAU, Navsari, Gujarat. His academic record is excellent and he qualified for the ASRB-ICAR NET in 2021 and 2023. He has completed a training, participated and present research paper in several national and international seminars, symposia and conferences.

Emerging Pattern in Plant Genomics and Marker Assisted Breeding

— *Volume 1* —

– Editors –

Lavudya Sampath

Tushar Arun Mohanty

Santhosh V.

Swapnil Baraskar

Prajapati Maulik R.

2025

Daya Publishing House®
A Division of

Astral International Pvt. Ltd.
New Delhi – 110 002

Published by : **Daya Publishing House**®
A Division of
Astral International Pvt. Ltd.
– ISO 9001:2015 Certified Company –
4736/23, Ansari Road, Darya Ganj
New Delhi-110 002
Ph. 011-43549197, 8130496929
E-mail: info@astralint.com
Website: www.astralint.com

Acknowledgement

The creation of "Emerging Pattern in Plant Genomics and Marker-Assisted Breeding Volume 1" has been a collaborative endeavor that would not have been possible without the support, dedication, and contributions of many individuals and organizations. We extend our heartfelt gratitude to those who have been instrumental in bringing this book to fruition. We would like to express our sincere appreciation to the authors of the individual chapters, whose expertise, research, and insights have enriched this volume. Your commitment to advancing the field of plant genomics and breeding is commendable. Our thanks also go to the peer reviewers who provided invaluable feedback and ensured the quality and accuracy of the content. Your rigorous examination of the chapters has enhanced the credibility of this work. We are grateful to the editorial and production teams who worked tirelessly to shape this book into its final form. Your professionalism, attention to detail, and dedication to excellence have been instrumental in the success of this project.

We acknowledge the academic institutions and research organizations that have supported our contributors in their scholarly pursuits. Your commitment to fostering a conducive environment for scientific exploration is appreciated.

Furthermore, we extend our thanks to the funding agencies and sponsors who have provided financial support for the research projects that underpin the chapters in this book. Your investment in scientific endeavors has a far-reaching impact on agriculture and food security.

To our colleagues, mentors, and peers in the field of plant genomics and breeding, we appreciate your camaraderie, intellectual exchange, and encouragement. Your passion for pushing the boundaries of knowledge continues to inspire us. Lastly, we express our gratitude to our families and loved ones for their unwavering support and understanding throughout the book's development. Your patience and encouragement have been a source of strength. This book is a testament to the collective effort of the scientific community and reflects our shared commitment to advancing the frontiers of plant genetics and crop improvement. Thank you all for your contributions, dedication, and enthusiasm.

Lavudya Sampath

Tushar Arun Mohanty

Santhosh V.

Swapnil Baraskar

Prajapati Maulik R.

Preface

In the ever-evolving world of agriculture and plant breeding, the power of genomics has unleashed a new era of possibilities. "Emerging Patterns in Plant Genomics and Marker-Assisted Breeding Volume 1" represents a comprehensive exploration of the cutting-edge research and innovative methodologies shaping the future of plant breeding. This volume is a testament to the dedication and curiosity of scientists, researchers, and practitioners in the field of plant genomics. With each passing day, our understanding of plant genetics deepens, opening up exciting avenues for enhancing crop productivity, resilience, and nutritional value. As we stand on the cusp of a green revolution driven by molecular insights, it becomes increasingly essential to gather and disseminate the latest knowledge.

The chapters in this volume cover a wide spectrum of topics, from gene isolation and functional analysis to the application of high-throughput phenotyping in modern plant breeding. Each chapter is a building block in the larger edifice of knowledge, contributing to our understanding of plant genomes and their intricate regulatory networks. We begin with a journey into the world of gene isolation and functional analysis, exploring the fundamental elements of plant genetics. Advancements in callus morphogenesis and gene transformation techniques pave the way for genome editing, offering unprecedented precision in plant breeding.

Molecular markers, such as SNPs and InDels, have revolutionized the field, and we delve into their genotyping methods and applications.

Transcriptomics, proteomics, and metabolomics provide insights into the molecular underpinnings of plant biology, while omics integration offers a holistic approach to breeding.Epigenomics and non-coding RNAs emerge as key players, influencing crop improvement strategies. The search for novel mutants and the principles of QTL mapping guide us toward harnessing novel traits for crop enhancement. Techniques like site-directed mutagenesis and haploid production expand our toolkit for crop improvement. High-throughput phenotyping and the role of bioinformatics in data analysis contribute to the ever-growing body of knowledge. This volume is a testament to the collaborative spirit of scientists across the globe who are dedicated to addressing the challenges of food security and sustainability. It is our hope that this compilation will serve as a valuable resource for researchers, educators, and students, fostering a deeper understanding of plant genomics and its applications in modern agriculture. As we embark on this journey through the emerging patterns in plant genomics and marker-assisted breeding, we invite you to explore the limitless potential of genomics in shaping the future of agriculture.

Lavudya Sampath

Tushar Arun Mohanty

Santhosh V.

Swapnil Baraskar

Prajapati Maulik R.

Contents

Chapter 1

Gene Isolation and Functional Analysis

Amita Ekka and Jhanendra Kumar Patel*

Ph.D. Scholar, Department of Genetics and Plant Breeding, IGKV, Raipur – 492 012, Chhattisgarh
**e-mail: amitaekka95@gmail.com*

ABSTRACT

In order to understand the molecular mechanisms behind numerous biological processes in plants, gene isolation and function study are essential. The approaches and developments in the area of gene isolation and function analysis are outlined in this review, with an emphasis on how they have enhanced our knowledge of plant biology. The entire process of gene includes recombinant DNA technology, Reverse transcriptase and PCR analysis mainly. The analysis becomes more complex as a result of the incorporation of computer methods for forecasting gene function and interactions. Along with case studies demonstrating effective gene separation and subsequent functional characterization, the role of model species and their significance in gene function research are highlighted. It is also emphasized how important gene function analysis is for crop improvement, stress tolerance, and sustainable agriculture. In conclusion, this study highlights the vital contribution that gene isolation and function analysis provide to the advancement of plant science, paving the way for creative approaches to crop improvement and tackling the world's agricultural problems. This chapter includes methods of gene isolation and function analysis of genes.

Keywords: Gene isolation, PCR, Crop improvement, Molecules.

INTRODUCTION

Gene, as we all know how gene function in inheritance, how they in-coded protein and their significance, but it doesn't deliver us the entire study, to understand the gene function in profundity gene isolation in its purest form is required. Once we get the gene isolation it can be studied in various ways like biochemical, molecular and biochemistry *Taq* and able to know the functions of particular gene which can be the game changer in future perspective.

We are all friendly with how genes work in heredity, how they encode proteins, and their chemical makeup. The isolation of a gene in a pure form is crucial to fully comprehending how a gene functions. The gene can be examined at multiple biochemical, molecular, and biochemistry levels once it has been isolated.

However, techniques for isolating genes were not created until the 1960s, and they were only useful for a small number of genes. With the development of molecular cloning or recombination DNA technology in the late 1970s, all of this altered. This method enables researchers to separate any gene from any individual from which entire DNA/RNA might be extracted.

As whole genomes are sequenced, the full potential to give access to all of an organism's genes is now being realized. After the invention of recombinant DNA technology, extensive research into individual DNA molecules produced several results, one of which was an isolation method for any DNA whose sequence is known. This technique is known as Polymerase Chain Reaction (PCR).

Methods for Gene Isolation

These techniques have transformed disciplines including genetics, genomics, biotechnology, and medicine and allowed for deep insights into the complexities of life. Recombinant DNA technique has its origins in bacterial genetics, like many other developments in molecular genetics.

The first gene to be extracted was a bacterial gene that bacteriophage could ingest. By separating these hybrid bacteriophages, the bacterial gene's DNA could be retrieved in a highly enriched form. Recombinant DNA technology operates on this fundamental tenet.

The first successful isolation of gene from crop plant can be attributed to the work of scientists who discovered and characterized the alcohol dehydrogenase (Adh) gene in maize (corn). In 1983, Stephen Linn and colleagues published a paper titled "Nucleotide sequence of a B2 repetitive element from the maize genome" in the journal "Nature." The *Adh* gene served as a model for studying gene structure, regulation, and evolution. This accomplishment paved the way for subsequent research in genetic engineering and plant molecular biology, enabling scientists to manipulate genes for various agricultural purposes.

Recombinant DNA Technology

Recombinant DNA

The chemical underpinning of life can now be understood as a result of James Watson and Francis Crick's 1953 discovery of the DNA structure. This discovery made it possible to study genes, the basic units of heredity in charge of transmitting features from one generation to the next. Scientists have developed a number of gene isolation techniques that enable the extraction, purification, and study of certain genes or DNA segments in an effort to uncover the mysteries contained in genes (Hardison *et al.*, 2023).

Restrictions enzymes (RE), which first became known in the 1970s, are proteinaceous in nature that cleaves DNA at particular recognition sequences. Specific DNA fragments can be produced by researchers using various enzymes, which can then be extracted and further, investigated. RE cuts the DNA at specified (4-6 bp, on average) sequence. This enables the DNA of interest to be present in specific places. The physiological purpose of restriction endonuclease is to operate as a component of the defence mechanism that keeps viruses and other organisms from invading bacteria.

In the Recombinant DNA Technology principle, the four steps are as follows:

1. Gene cloning and Recombinant DNA development. 2. Vector entry into the host. 3. Picking a host cell that has been transformed. 4. The gene that was introduced is translated and is translated.

Nuclease enzymes are of two types:

1. **Exonucleases**: These type of enzymes cleave the nucleotides from the ends of DNA. Restriction exonucleases are primarily responsible for hydrolysis of the terminal nucleotides from the terminal end of DNA or RNA molecule either from 5′ to 3′ direction or 3′ to 5′ direction; for example- exonuclease I, exonuclease II, *etc.*

2. **Endonucleases:** These enzymes cause DNA at particular locations to be cut. It identify the particular base sequences (restriction sites) within DNA or RNA molecule and catalyze the cleavage of internal phosphodiester bond; *e.g.* EcoRI, Hind III, BamHI, *etc.*

There are mainly 3 types of restriction endonuclease which are listed in Table 1.1.

They are members of the group of enzymes known as nucleases and are also referred to as molecular scissors. Hamilton Smith, Werner Arber, and Daniel Nathans shared the Nobel Prize for discovering restriction endonuclease. The first restriction enzyme to be identified was **Hind II.** It came from the *Haemophilus influenza* bacteria. Many microbial species have been found to manufacture hundreds of restriction enzymes. They each have a unique limitation site which is represented in Table 1.2.

Table 1.1: Types of Restriction Endonuclease

Properties	Type I RE	Type II RE	Type III RE
Abundance	Less common than type II	More common	Rare
Recognition site	Cut both strands at a non-specific location >1000bp away from recognition site	Cut both strands at a specific usually palindromic recognition site (4-8 bp)	Cleavage of one strand only 24-26bp downstream of the 3' recognition site
Restriction and modification	Single multifunctional enzyme	Separate nuclease and methylase	Separate enzymes sharing a common subunit
Nuclease subunit structure	Heterotrimer	Homodimer	Heterodimer
Co-factors	ATP, Mg^{2+}, SAM	Mg^{2+}	Mg^{2+}, SAM
DNA cleavage requirement	Two recognition site in any orientation	Single recognition site	Two recognition sites in a head-to-head orientation
Enzymatic turnover	No	Yes	Yes
DNA translocation	Yes	No	No
Site of methylation	At recognition site	At recognition site	At recognition site

Table 1.2: List of Enzymes Use for Gene Isolation

Name of the Enzyme	Source	Recognition Site and Cleavage Site	Nature of cut ends
Eco RI	*E. coli* RY13	5'-G|AATTC-3'3'-CTTAA|G-5'	Sticky
Hind III	*Haemophilius influenza* Rd	5'-A|AGCTT-3'3'-TTCGA|A-5'	Sticky
Bam HI	*Bacillus amyloliquifaciens*	5'-G|GATCC|-3'3'-CCTAG|G-5'	Sticky
Bal I	*Brevibacterium albidum*	5'-TGG|CCA-3'3'-ACC|GGT-5'	Blunt
Hae III	*Haemophilus aegipitus*	5'-GG|CC-3'3'-CC|GG-5'	Blunt
Sma I	*Serratia marcescens*	5'-CCC|GGG-3'3'-GGG|CCC-5'	Blunt

Reverse Transcription

It is a basic molecular biology technique that involves the conversion of RNA into complementary DNA (cDNA). This process plays a significant role in various areas of research, including gene expression analysis, cloning, and the study of retroviruses (Maga, 2013).

Principle of Reverse Transcription

Enzyme Reverse transcriptase catalyzes the process of reverse transcription, which is typically derived from retroviruses or synthesized in the laboratory. Reverse transcriptase synthesizes complementary DNA strand from template to a, resulting in the formation of cDNA.

Applications

Gene Expression Analysis: Reverse transcription is a pivotal step in quantitative PCR (qPCR) and reverse transcription-quantitative PCR (RT-qPCR) assays. It allows scientists to convert RNA molecules (mRNA) into c-DNA, which can be amplified by PCR and quantified.

Cloning: Reverse transcription is frequently used in c-DNA library construction. This involves creating a set of cDNA molecules representing the genes expressed in specific tissue or under certain conditions.

Viral Research: The replication cycle of retroviruses involves reverse transcription. This step is important for the creation of DNA from the viral RNA genome, which can then be incorporate into the host cell's genome.

Process: The process of reverse transcription involves several key steps:

Primer Binding: A short DNA primer is annealed to RNA template. This primer provides a initial point for reverse transcriptase to initiate cDNA synthesis.

Reverse Transcription: Reverse transcriptase synthesizes a complementary DNA strand using the RNA template and the DNA primer.

RNA Degradation: An RNase H activity associated with reverse transcriptases which degrade the RNA template.

Second Strand Synthesis: Reverse transcriptase may have an associated DNA polymerase that synthesizes the second DNA strand, completing the cDNA synthesis.

PCR (Polymerase Chain Reaction): An effective technique for in vitro DNA molecule multiplication in the biological sciences is the PCR. As a result, the method has many uses in both basic and applied research in medicine, agriculture, the environment, and the bio-industry.

In present molecular biology, one of the most widely used techniques is the PCR. The method was created in year 1984 by Nobel laureate Kary Mullis. A target DNA sequence is multiplied extremely precisely in vitro, making it simple to handle and investigate using standard techniques of molecular biology. In 1988, the first PCR machine entered the market. PCR-based methodologies led to the Human Genome Project. PCR methods have undergone several variations over the past few decades due to their vast range of applications.

PCR Fundamentals

The PCR procedure entails 20 to 40 cycles of repetitive temperature changes, with each cycle typically including three distinct temperature variations (Mullis *et al.*, 1990; Rajalaxmi *et al.*, 2017).

1. **Denaturation** 94–96°C,
2. **Primer annealing** (depending on primer): 45–60°C,
3. **Primer extension/Polymerization**: usually 72°C

The temperature step known as "hold" at which product extension is carried out occurs frequently at the beginning and final stages of the cycle phases at (>90) and (~72), respectively. Prior to analysis, the finished product is stored at 4°C. Most PCR techniques are normally capable of amplifying DNA fragments up to 10kb. However, certain methods can multiply up to 40 kb.

The PCR's Component Parts

The components, chemicals, and conditions needed to set up a PCR are listed below.

DNA Template

The template DNA (target DNA) is the ds-DNA molecule that PCR amplifies. During the process of PCR, the template specifies the order in which additional nucleotides (dNTPs) are added (Bell *et al.*, 2008). By cyclically altering the temperature, this procedure is carried out in vitro and allows for the separation of ds-DNA, primer hybridization, and polymerization. As long as the target sequence is present, DNA template can be exploiting as a PCR template. We need to know something about the target DNA sequence in order to construct PCR primers (Ye *et al.*, 2012).

Primers

Primers are polynucleotides of varying length that are typically produced commercially as ss-DNA molecules (Zaider *et al.*, 2011; Joshi *et al.*, 2010). The free 32-OH hydroxyl group, commonly known as the 32 end, is present in these short polynucleotide DNA strands. The DNA polymerase requires the free 32 hydroxyl group on the primer in order to incorporate additional nucleotides throughout the polymerization process and so create a new complementary strand (Olerup *et al.*, 1992). The separation of two complementary DNA strands (Denaturation) is necessary for the binding of DNA primer to the target. 2 primers are required for PCR in order to enhance both strands of the template: one primer for the sense strand, also known as the forward primer, and another primer for complementary strand, also known as the antisense strand, also known as the reverse primer. As a result, the 52 ends of the sense and antisense strands are bound by both primers. In order to appropriately identify their assigned target complementary areas, primer length is crucial.

DNA Polymerase

The 32 end of the template is where DNA polymerase begins to create new

DNA. The primers attach to the 32 ends of the two template stands, which are subsequently expanded by DNA polymerase. Taq DNA polymerase, which was derived from the thermophilic bacteria *Taq*, is the most widely used DNA polymerase. Taq-polymerase increases the length of the DNA chain by averaging 1.0 kb per minute, with an enzyme half-life of 40 minutes at 95 °C. As an alternative, because *Taq* DNA polymerase lacks 32-52 exonuclease activity (proof-reading), *Pyrococcus furiosus* (*Pfu*) DNA polymerase is also extensively utilised. By removing wrongly inserted nucleotides during polymerization, proofreading enables *Pfu* to create new DNA with the fewest possible mistakes.

Nucleotides

To create new DNA strands, PCR needs four distinct deoxynucleoside triphosphates, or dNTPs: adenine (A), guanine (G), cytosine (C), and thymine (T). In reaction mixture, the deoxynucleoside triphosphates are typically given at a conc. of 200μM. These four deoxynucleoside triphosphates must all be present in equal concentrations in mixture; otherwise, the *Taq*-DNA polymerase may incorrectly incorporate nucleotides.

Buffer Solution

The objective of PCR buffer solution is to give the DNA polymerase an appropriate environment and components for maximum activity and stability (Hawanget *et al.*, 2003). The buffers frequently include KCl, Tris-HCl, and occasionally $MgCl_2$. The components listed in Table 1.3 are among the Taq formulation-specific PCR buffers that are frequently offered in 10x concentrations.

Table 1.3: Component of Polymerase Chain Reaction and their Functions

Component	Functions
100 mM Tris-HCl (pH - 8.8 at 25 °C)	Maintain the pH of reaction
500 mM KCl	Stabilize the annealing of primer-template
15 mM **$MgCl_2$**	DNA polymerase Co-factor
0.8 per cent (v/v) Nonidet P40 (Optional)	Suppresses formation of secondary structure

(Thermo Fisher Scientific™B16: https://www.thermofisher.com/order/catalog/product/B16)

Thermal Cycler

The PCR mixtures are quickly heated and cooled using a thermal cycler, also known as a thermocycler, which cycles them through the 3 temperature steps of PCR. The modern types of thermocyclers are also equipped with heated covers that escape reaction mixture condensation while PCR is being run. In earlier thermocyclers without this feature, by placing oil/wax balls on the PCR mixture's surface, the evaporation was stopped.

Process

Denaturation, annealing, and polymerization are the three main processes in each round of PCR, and this repetition for 30-40 cycles using a thermo cycler. For DNA synthesis, DNA polymerase is used, the primers melting temperature, and the concentration of the reagents employed, such as divalent ions and dNTPs, are only a few of the factors that affect the temperature range and length of each round (Figure 1.1). The primer specific nucleotide sequence and length affect its melting temperature.

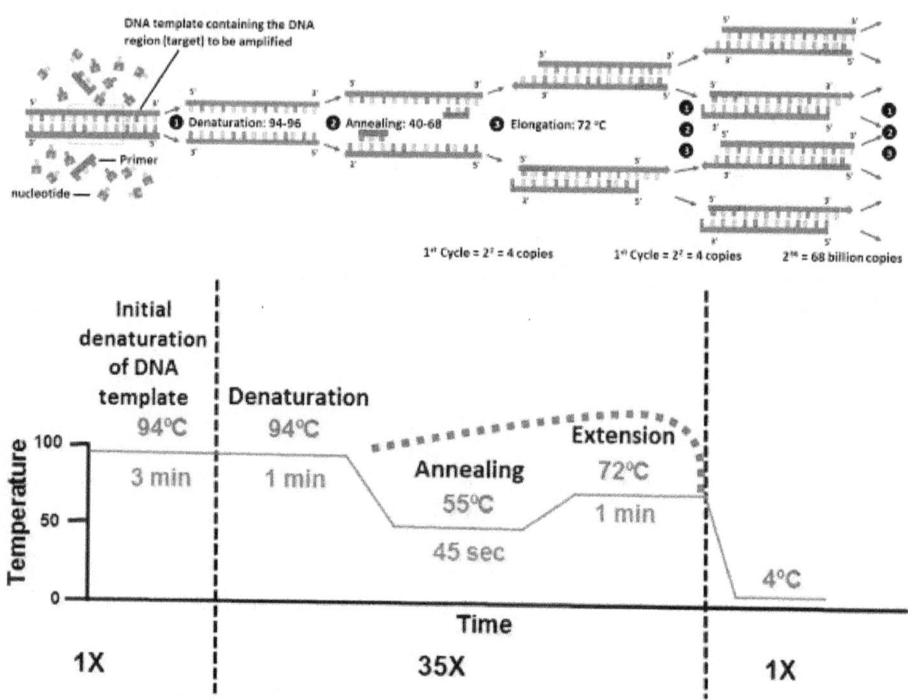

Figure 1.1: Process of PCR Multiplication of DNA.

Application of PCR

The PCR method and its numerous sophisticated versions function as strong tools with specific applications that were previously unthinkable by the researcher community (Green *et al.*, 2012). Analysis of Molecular genetics approaches, which includes the quick the infectious diseases diagnosis and paternity detection, as well as gene expressions analysis, whole genome sequence, in recombinant techniques, and other scientific advancements were all made possible by this adaptable technique (Valones *et al.*, 2009). It allows for the in-vitro synthesis of DNAs/RNAs that can specifically repeat a DNA segment in a semiconservative manner.

Functions of Isolated Gene

Gel Electrophoresis

DNA fragments are seperated using the gel electrophoresis method according to their size and charge. This method involves placing DNA samples on a gel matrix and applying an electric pulse. Larger DNA fragments migrate slower than smaller ones, resulting in distinct bands that can be visualized under ultraviolet light. This technique aids in identifying DNA size, quantifying DNA, and analyzing genetic mutations.

Hybridization Techniques

These methods involve the use of complementary DNA or RNA probes to locate specific DNA sequences. Southern blotting, for example, identifies target DNA fragments by hybridizing them with labeled probes. Northern blotting does the same for RNA. In situ hybridization allows researchers to visualize the location of specific genes within cells or tissues.

DNA Cloning

DNA cloning involves isolating a specific gene or DNA fragment and then inserting it into a vector, such as a bacterial plasmid. This creates a recombinant DNA fragments that are replicated within host cells. This method is crucial for producing large quantities of a gene/DNA of interest and studying its function.

Genomic Libraries

A genomic library is a collection of cloned DNA fragments representing an organism's entire genome. These libraries serve as valuable resources for studying gene function and detecting specific genes which is accountable for various traits or diseases. c-DNA libraries, contain only the expressed genes and are useful for identification of gene expression patterns.

PCR-Based Techniques

Beyond basic PCR, various PCR-based techniques have emerged. Reverse Transcription PCR (RT-PCR) is utilized to multiply RNA by first converting it to complementary DNA using the enzyme reverse transcriptase. Quantitative PCR (qPCR) allows for the quantification of gene expression levels, while RT-qPCR combines reverse transcription and qPCR to analyze RNA expression.

Next-Generation Sequencing (NGS)

Next-Generation Sequencing technology has reshaped gen-omics by enabling the rapid sequencing of DNA at a massive scale. Methods like Illumina sequencing can sequence billions of DNA fragments in parallel, providing insights into an individual's entire genome or the genome of an entire population.

CRISPR-Cas9

The development of the CRISPR-Cas9 system has revolutionized genetic engineering and gene editing. This technology permits for exact modification of DNA sequences, enabling researchers to edit genes with unprecedented accuracy. CRISPR-Cas9 has vast applications, from creating genetically modified organisms to potential therapeutic interventions for genetic disorders.

Table 1.4: List of some Isolated Genes along with their Particular Trait and References

Isolated Gene	Trait	References	Remarks
Adh$_r$	The activity of alcohol dehydrogenase in the scutellum of maize	Efron, 1970	This gene is not allelic to the Adh1 locus, which specifies the charge of the enzyme molecule
Ac-Ds element	Homology of Ac-Ds element, they are transposable within a genome.	Fedoroff *et al.*, 1983	Isolated from Maize genome.
TAG1 gene (Tomato AGAMOUS gene)	Differentiate into stamens and carpels and later becomes restricted to specific cell types within these organs.	Pnueli *et al.*, 1994	The AGAMOUS gene of *Arabidopsis* is necessary for the proper development of stamens and carpels and the prevention of indeterminate growth of the floral meristem.
LiYAB1	A *YABBY* gene family, from lily (*Lilium longiflorum*), is a plant-specific transcriptional factor.	Wang *et al.*, 2009	*LiYAB1* is expressed strongly in the carpels of the lily flower and weakly in the leaves, which is similar to *DL* in rice
Deacetylvin-doline-4-*O*-acetyl-transferase (DAT)	The key enzyme for biosynthesis of vindoline in Madagascar periwinkle (*Catharanthus roseus*)	Wang *et al.*, 2010	Approx. 1773 bp genomic DNA fragment of DAT promoter contained various potential regulatory components which were involved in the regulation of gene expression.
CHI **(844 bp in length)**	The essential enzymes in the anthocyanin biosynthetic pathway catalyzing the stereospecific isomerization of chalcones into their corresponding (2S)-flavanones.	Guo *et al.*, 2015	Isolated from purple-fleshed sweet potato [*Ipomoea batatas* (L.) Lam] cv. *Yamakawa murasaki*
VyMYB24	Regulatory role in the phenylpropanoid pathway including proanthocyanidin, anthocyanin, and flavonoid biosynthesis	Zhu *et al.*, 2022	Isolated from a high drought-tolerant Chinese wild Vitis species *V. yanshanesis*.

Isolated Gene	Trait	References	Remarks
RF2 Gene	RF2 family genes responded to salt stress and drought stress in cotton.	Gu *et al.,* 2023	Silencing the GhRF2-32 gene showed less leaf wilting and increased total antioxidant capacity under drought and salt stress, decreased malondialdehyde content, and increased drought and salt tolerance
SmMYB39	Regulates salt stress tolerance of Eggplant (*Solanum melongena* L.).	Jiang *et al.,* 2023	Silencing of *SmMYB39* led to reduced salt stress tolerance in eggplant
Eight A1 (*Asa DREB1.1–1.8*) and eight A2 (*Asa DREB2.1–2.8*) genes	*AsaDREB2* genes are affected by low temperatures and *Fusarium* infection, suggesting their contribution to garlic adaptive responses. *AsaDREB2.4* could be a pseudogene, whereas the pseudogenic status of *AsaDREB2.6* and *2.7* may be genotype-specific.	Filyushin *et al.,* 2023	Hormone- and stress-responsive *cis*-regulatory elements.
SrMYB1	Direct transcriptional repressor of SrUGT76G1 in *Stevia rebaudiana*	Zhang *et al.,* 2023	SrUGT76G1 is key gene to the biosynthesis of important steviol glycosides (SGs).
FvICE1	Positively regulator of cold and drought resistances in *Fragaria vesca* through over-expression and CRISPR/Cas9 technologies	Han *et al.,* 2023	A member of the bHLH TF family, with a length of 1608 bp
lncRNA77580 long (non-coding RNAs)	Regulates Drought and Salinity Stress Responses in Soybean	Chen *et al.,* 2023	Overexpression of lncRNA77580 enhances drought tolerance, increased sensitivity to high salinity at the seedling stage and underwater deficit at the reproductive stage, and overexpression increases the seed yield by increasing the seed number/plant.
com58276	The function of *com58276* is conserved across species, improving the tolerance of cotton to salt and low temperature, and demonstrating its applicability to improve plant resistance to environmental change.	Pu *et al.,* 2023	Isolated from the desert plant *Caragana kosinski*.

Isolated Gene	Trait	References	Remarks
CmMYB3	Regulates flavonol biosynthesis in *Arabidopsis thaliana*	Yang *et al.*, 2023	Isolated from the *Chrysanthemum morifolium*. In overexpression of CmMYB3 in *Nicotiana benthamiana* and *Arabidopsis thaliana*, he flavinol contents in plants were increased, and the expression of AtCHS, AtCHI, and AtFLS genes in *Arabidopsis thaliana* was also improved.

Table 1.5: Application and Functions of Isolated Gene

Application	Description	References
Meta-genomics	Gene-targeted metagenomics combines PCR with metagenomics to identify rarest members of a sampled community and rare genes in the community members.	Rosen *et al.*, 2012 and Lee *et al.*, 2014
Site-directed mutagenesis	PCR-based approaches are commonly used to insert mutations (deletions, additions, and substitutions) at specific locations in a gene to study role of specific amino acids in the structure and function of proteins.	Rouached *et al.*, 2010 and Edelheit *et al.*, 2009
Personalized medicine	PCR technologies are employed in pharmaco-genomics and pharmaco-genetics to track genetic markers that determine the response of individuals to treatments and are used to design tailor-made drugs and to prescribe drugs in effective doses	Dwivedi *et al.*, 2017 and Chu *et al.*, 2012
Forensics sciences	The power of PCR is employed to amplify poor quality and quantity DNA samples from crime scenes and make them reliably analysable.	Green *et al.*, 2012 and Lakshmi *et al.*, 2011
DNA profiling	DNA profiling methods utilize PCR-based approaches to exploit the polymorphic nature of DNA (SNPs, DNA repeats, *etc.*) to study the structure and diversity ecological communities, phylogeny, and population genetics.	Epplen *et al.*, 2012
Gene expression profiling	Reverse-transcriptase PCR and qPCR are routinely employed to profile the expression of genes and to validate transcriptome profiles generated through techniques like microarray and RNA-seq	Mooney *et al.*, 2012 andSheibani *et al.*, 2015
Identifying medicinal plants	PCR-based DNA barcoding is a tool that utilizes specific DNA sequences to rapidly and accurately identify medicinal plants species from other morphologically similar plants. This approach is also used by ecologists and conservation biologists to identifying endangered and new species.	Krees *et al.*, 2005

Application	Description	References
Detecting GMO	PCR techniques are used to quickly and reliably track the presence of genetically modified organism in food and feed to ensure their regulation and protection of consumer rights	Rabiei *et al.*, 2013

Conclusions

In conclusion, gene isolation methods have played a pivotal role in unravelling the intricacies of life's blueprint, the DNA. These methods, from traditional techniques like gel electrophoresis and restriction enzyme digestion to cutting-edge technologies like CRISPR-Cas9 and NGS, have not only enabled scientists to study genes but have also paved the way for breakthroughs in genetics, medicine, and biotechnology. As our understanding of genes continues to deepen, these methods will undoubtedly evolve, ushering in new possibilities for uncovering the mysteries of life itself.

REFERENCES

Bell LE. Cooling, heating, generating power, and recovering waste heat with thermoelectric systems. Science. 2008; 321(5895): 1457-1461.

Bermingham N, Luettich K. Polymerase chain reaction and its applications. Current Diagnostic Pathology. 2003; 9(3): 159-164.

Chen X, Jiang X, Niu F, Sun X, Hu Z, Gao F *et al.*, Overexpression of lncRNA77580 Regulates Drought and Salinity Stress Responses in Soybean. Plants. 2023; 12: 181.

Cho SH, Jeon J, Kim SI. Personalized medicine in breast cancer: A systematic review. Journal of Breast Cancer. 2012;15(3): 265-272

Cortelli S, Jorge AO, Querido SM, Cortelli JR. PCR and culture in the sub gingival detection of Actinobacillus actinomycetemcomitans: A comparative study. Brazilian Dental Science. 2010; 6(2).

Dubey VK. Proteomics and Genomics. Lecture 37: Polymerase Chain Reaction. Available from: http://nptel.ac.in/courses/102103017/module37/lec37_slide1.htm (Accessed: 24 May 2018)

Dwivedi S, Purohit P, Misra R, Pareek P, Goel A, Khattri S, *et al.*, Diseases and molecular diagnostics: A step closer to precision medicine. Indian Journal of Clinical Biochemistry. 2017; 32(4): 374-398

Edelheit O, Hanukoglu A, Hanukoglu I. Simple and efficient site-directed mutagenesis using two single-primer reactions in parallel to generate mutants for protein structure function studies. BMC Biotechnology. 2009; 9(1): 61.

Efron Y. Alcohol dehydrogenase in maize: genetic control of enzyme activity. Science. 1970; 13: 170(3959): 751-3.

Epplen J, Lubjuhnn T. DNA Profiling and DNA Fingerprinting. Springer Science and Business Media; 11 Dec 2012.

Fedoroff N, Wessler S, Shure M. Isolation of the transposable maize controlling elements Ac and Ds. Cell. 1983; 35(1): 235-42.

Filyushin MA, Anisimova OK, Shchennikova AV and Kochieva EZ. DREB1 and DREB2 Genes in Garlic (*Allium sativum* L.): Genome-Wide Identification, Characterization, and Stress Response. Plants. 2023; 12: 2538.

G. Maga, in Brenner's Encyclopedia of Genetics (Second Edition), 2013. Pp: 222-223. https://doi.org/10.1016/B978-0-12-374984-0.01326-7

Green MR, Sambrook J, Sambrook J. Molecular Cloning: A Laboratory Manual. 4th ed. New York: Cold Spring Harbor Laboratory Press; 2012.

Gu H, Zhao Z, Wei Y, Li P, Lu Q, Liu Y *et al.*, Genome-Wide Identification and Functional Analysis of RF2 Gene Family and the Critical Role of GhRF2-32 in Response to Drought Stress in Cotton. Plants. 2023: 12: 2613.

Guo J, Zhou W, Lu Z, Li H, Li H and Gao F. Isolation and Functional Analysis of Chalcone Isomerase Gene from Purple-Fleshed Sweet Potato. Plant Mol Biol Rep. 2015; 33: 1451–1463.

Han J, Li X, Li W, Yang Q, Li Z, Cheng Z *et al.*, Isolation and preliminary functional analysis of FvICE1, involved in cold and drought tolerance in *Fragaria vesca* through over-expression and CRISPR/Cas9 technologies. Plant Physiology and Biochemistry. 2023; 196: 270-280.

Hwang IT, Kim YJ, Kim SH, Kwak CI, Gu YY, Chun JY. Annealing control primer system for improving specificity of PCR amplification. Bio Techniques. 2003; 35(6): 1180-1190.

Jiang Z, Shen L, He J, Du L, Xia X, Zhang L and Yang X. Functional Analysis of SmMYB39 in Salt Stress Tolerance of Eggplant (*Solanum melongena* L.). Horticulturae. 2023; 9: 848.

Joshi M, Deshpande JD. Polymerase chain reaction: Methods, principles and application. International Journal of Biomedical Research. 2010; 2(1): 81-97

Kress WJ, Erickson DL. DNA barcodes: genes, genomics, and bioinformatics. Proceedings of the National Academy of Sciences. 26 Feb 2008; 105(8): 2761-2762

Lakshmi V, Sudha T, Rakhi D, Anilkumar G, Dandona L. Application of polymerase chain reaction to detect HIV-1 DNA in pools of dried blood spots. Indian Journal of Microbiology. 2011; 51(2): 147-152.

Lakshmi V, Sudha T, Rakhi D, Anilkumar G, Dandona L. Application of polymerase chain reaction to detect HIV-1 DNA in pools of dried blood spots. Indian Journal of Microbiology. 2011; 51(2): 147-152.

Lee S, Cantarel B, Henrissat B, Gevers D, Birren BW, Huttenhower C, *et al.,* Gene-targeted metagenomic analysis of glucan-branching enzyme gene profiles among human and animal fecal microbiota. The ISME Journal. 2014; 8(3): 493-503

Mooney C, Raoof R, El-Naggar H, Sanz-Rodriguez A, Jimenez-Mateos EM, Henshall DC. High throughput qPCR expression profiling of circulating microRNAs reveals minimal sex-and sample timing-related variation in plasma of healthy volunteers. PLoS ONE. 23 Dec 2015;10(12): e0145316

Mullis KB. The unusual origin of the polymerase chain reaction. Scientific American. 1990; 4: 56-65.

Olerup O, Zetterquist H. HLA-DR typing by PCR amplification with sequence-specific primers (PCR SSP) in 2 hours: An alternative to serological DR typing in clinical practice including donor-recipient matching in cadaveric transplantation. HLA. 1992; 39(5): 225-235

Patel SV, Bosamia TC, Bhalani HN, Singh P, Kumar A. Polymerase chain reaction (PCR). Agrbios. 2015; 13(9): 10-12.

Pnueli L, Hareven D, Rounsley SD, Yanofsky MF and Lifschitz E. Isolation of the tomato AGAMOUS gene TAG1 and analysis of its homeotic role in transgenic plants. Plant Cell. 1994; 6(2): 163-73.

Pu Y, Wang P, Xu J, Yang Y, Zhou T, Zheng K *et al.,* Over-expression of the *Caragana korshinskii* com58276 Gene Enhances Tolerance to Drought in Cotton (*Gossypium hirsutum* L.). Plants. 2023; 12: 1069.

Rabiei M, Mehdizadeh M, Rastegar H, Vahidi H, Alebouyeh M. Detection of genetically modified maize in processed foods sold commercially in Iran by qualitative PCR. Iranian Journal of Pharmaceutical Research. 2013; 12(1): 25.

Rajalakshmi S. Different types of PCR techniques and its applications. International Journal of Pharmaceutical, Chemical and Biological Sciences. 2017; 7(3): 285-292.

Rosen MJ, Callahan BJ, Fisher DS, Holmes SP. Denoising PCR-amplified metagenome data. BMC Bioinformatics. 2012; 13(1): 283.

Rouached H. Efficient procedure for site-directed mutagenesis mediated by PCR insertion of a novel restriction site. Plant Signaling and Behavior. 2010; 5(12): 1547-1548

Sárosi I, Gerald E, Girish VN. Polymerase chain reaction and its application. Orvosi Hetilap. 1992; 133: 16-21

Shaheen S, Mohammad A, Sikandar S and Imran A. Genetic Engineering - A Glimpse of Techniques and Applications. 2020. pp: 1-20.

Sheibani-Tezerji R, Rattei T, Sessitsch A, Trognitz F, Mitter B. Transcriptome profiling of the endophyte Burkholderia phytofirmans PsJN indicates sensing of the plant environment and drought stress. MBio. 30 Oct 2015;6(5): e00621-e00615

Valones MAA, Guimarães RL, Brandão LAC, de Souza PRE, de Albuquerque Tavares Carvalho A, Crovela S. Principles and applications of polymerase chain reaction in medical diagnostic fields: A review. Brazilian Journal of Microbiology. 2009; 40(1): 1-11.

Wang A, Tang J, Li D, Chen C, Zhao X and Zhu L. Isolation and functional analysis of *LiYAB1*, a *YABBY* family gene, from lily (*Lilium longiflorum*). 2009; 166(9): 988-995.

Wang Q, Yuan F, Pan Q, Li M, Wang G, Zhao J and Tang K. Isolation and functional analysis of the *Catharanthus roseus* deacetylvindoline-4-O-acetyltransferase gene promoter. Plant cell reports. 2010; 29; 185-192.

Yang F, Wang T, Guo Q, Zou Q and Yu S. The CmMYB3 transcription factors isolated from the *Chrysanthemum morifolium* regulate flavonol biosynthesis in *Arabidopsis thaliana*. Plant Cell Reports. 2023; 42(4): 791-803.

Ye J, Coulouris G, Zaretskaya I, Cutcutache I, Rozen S, Madden TL. Primer-BLAST: A tool to design target specific primers for polymerase chain reaction. BMC Bioinformatics. 2012; 13(1): 13.

Zádor E. The polymerase chain reaction. Theoretical course: Basic biochemical methods and ischemic heart models. Supported by: HURO/0901/069/2.3.1 HU-RO DOCS. 2011. Available from: http://www.academia.edu/35361310.

Zhang T, Zhang Y, Sun Y, Xu X, Wang Y, Chong X *et al.*, Isolation and functional analysis of SrMYB1, a direct transcriptional repressor of SrUGT76G1 in *Stevia rebaudiana*. Journal of Integrative Agriculture. 2023; 22(4): 1058–1067.

Zhu Z, Quan R, Chen G, Yu G, Li X, Han Z *et al.*, An R2R3-MYB transcription factor VyMYB24, isolated from wild grape *Vitis yanshanesis* JX Chen., regulates the plant development and confers the tolerance to drought. Front. Plant Sci. 2022; 13: 966641.

Chapter 2

Advances in Callus Morphogenesis and Gene Transformation Techniques

Shobica Priya Ramasamy[*1], *Monika S.*[2], *Shubham Rajaram Salunkhe*[3] *and Anushya Ravi*[1]

[1]*Ph.D. Scholar, Centre for Plant Breeding and Genetics, Tamil Nadu Agricultural University, Coimbatore – 641 003*
[2]*Ph.D. Scholar, Department of Genetics and Plant Breeding, Anbil Dharmalingam Agricultural College and Research Institute, TNAU, Trichy – 620 009*
[3]*Ph.D. Scholar, Department of Plant Biotechnology, Centre for Plant Molecular Biology and Biotechnology, Tamil Nadu Agricultural University, Coimbatore – 641 003*
**e-mail: shobicaramasamy@gmail.com*

ABSTRACT

Plant tissue culture has revolutionized agricultural and biotechnological practices, enabling the manipulation of plant cells to achieve various objectives. This abstract delves into recent advancements in callus morphogenesis and gene transformation techniques, highlighting their significance in modern plant biotechnology. Callus formation, a critical aspect of plant tissue culture, involves the cultivation of undifferentiated cells, offering a versatile platform for applications like crop improvement and secondary metabolite production. The understanding of callus morphogenesis has advanced significantly, with a focus on the interplay of hormonal cues, genetic factors, and environmental conditions that govern cell fate determination within callus tissues. Gene transformation techniques further amplify the potential of callus cultures by enabling the introduction of desirable traits. Innovations such as Agrobacterium-mediated transformation, particle bombardment, and electroporation have expanded the repertoire of methods for genetic modification. Notably, the integration of advanced genome editing tools like CRISPR/Cas9 has revolutionized precision genetic engineering within callus cultures, accelerating the development of plants with targeted improvements. The practical implications of these advancements are substantial. Callus-based strategies contribute to crop enhancement efforts, facilitating the development of plants with heightened resistance to diseases, tolerance to environmental stresses, and increased yields. Moreover, callus cultures serve as bioreactors for the production of valuable secondary metabolites and bioactive compounds, opening avenues for pharmaceutical and industrial applications. In conclusion, the progress in callus morphogenesis and gene transformation techniques marks a pivotal juncture in plant biotechnology.

Keywords: *Callus culture, Indirect and Direct gene transformation, Morphogenesis.*

INTRODUCTION

Callus morphogenesis is a complex process that lies at the core of plant tissue culture and regeneration. This phenomenon, characterized by the de-differentiation and subsequent re-differentiation of plant cells, has captured the attention of researchers for decades due to its potential applications in agriculture, horticulture, and biotechnology. Plant morphogenesis is a process controlled by hormones and regulated by complex network of genes resulting in spatial and temporal development of tissues, organs and embryos (de Almeida *et al.*, 2015). However, plant morphogenesis occurs throughout the process of ontogenesis by the constant functioning of vegetative and floral meristems (Gaarslev *et al.*, 2021). Representatives of numerous plant groups have been found to possess the ability to develop calli in vitro. Numerous researchers have demonstrated that different vegetative, generative, and embryonic organs can be used as explants for callus initiation. Several reports on in vitro callus development and the morphogenesis processes contribute to the creation of regenerants *i.e.* regenerated plants. Particularly notable progress has been made in identifying the molecular features of these processes, namely the involvement of a number of distinct gene systems in them (Wan *et at.*, 2023). Biotechnologies for the bulk production of economically important plants have been developed by using in vitro callus cultures.

In the realm of biology, the ability to manipulate genes has emerged as one of the most transformative achievements of modern science (Zhang *et al.*, 2019). Gene transformation, a revolutionary technique, has transcended traditional boundaries, offering the remarkable capability to introduce foreign genetic material into the genomes of organisms, reshaping their genetic makeup and influencing their traits (Ochman *et al.*, 2000). This powerful tool has not only revolutionized our understanding of genetics but has also opened avenues for groundbreaking applications in various fields, from medicine and agriculture to biotechnology and fundamental research. The journey of gene transformation techniques has evolved alongside our deepening understanding of genetics, molecular biology, and cellular physiology. Early methods relied on biological processes, such as viral infection or bacterial conjugation, to transfer genetic material between organisms (Arber, 2014). The advent of recombinant DNA technology and the development of molecular tools paved the way for more precise and controlled methods of gene transformation.

Over the years, a myriad of innovative techniques has emerged, each offering distinct advantages based on the characteristics of the target organism, the intended application, and the desired level of precision. From the pioneering use of *Agrobacterium tumefaciens* in plants to the recent advancements in CRISPR/Cas9-based genome editing, gene transformation methods have become increasingly sophisticated, enabling targeted modifications with unprecedented accuracy (Mahfouz *et al.*, 2014). In agriculture, genetically modified organisms (GMOs) offer solutions for crop yield improvement, pest resistance, and nutrient enhancement (Gbashi *et al.*, 2021). The biotechnology industry harnesses gene transformation to produce valuable pharmaceuticals, enzymes, and biofuels (Singh, 2011). From precision genome editing and synthetic biology to personalized medicine and environmental remediation, the ongoing evolution of gene transformation techniques promises to reshape how we interact with and manipulate the genetic fabric of life (Kelley *et al.*, 2014). This chapter attempts to provide a comprehensive overview on callus morphogenesis and gene transformation procedures, including their characteristics, underlying factors, and prospective implications.

Morphogenesis

The term morphogenesis is derived from Greek word meaning "transform". Morphogenesis is the process through which the cells that make up an organism's body, or portions thereof, multiply and differentiate without producing new tissue from previous material. This idea is most commonly used for describing plant development, particularly the formation of leaves and flowers. Plant formation and growth commence when a cell is fertilized by sperm, initiating embryogenesis as the initial phase. Afterward, cell division results in the formation of a seed, with adhering cells that gradually transform into a seedling. The term "monocot" refers to a plant that originates from a seed. As seedlings progress, they develop stems, roots, and leaves, a process termed morphogenesis. Plant morphogenesis primarily occurs due to differential growth. The presence of permanent embryonic tissue results in varying morphogenetic potential, greatly influenced by the environment. This potential continues to generate new plant organs throughout the plant's lifespan.

Plant Morphogenesis

Plant morphogenesis refers to the process through which plants develop their various structures and forms. There are several forms of plant morphogenesis that describe different aspects of plant growth and development (Sattler, 1992). Plant morphogenesis encompasses the multifaceted processes guiding the development of plants' diverse structures. These encompass embryogenesis, wherein a zygote transforms into a multicellular embryo, setting the groundwork for growth (Guo *et al.*, 2022). Shoot morphogenesis oversees above-ground structure formation, driven by shoot apical meristems fostering stem, leaf, flower, and fruit emergence. Root morphogenesis manages subterranean structure growth *via* root apical meristems, sustaining continuous growth and new tissue generation. Vascular morphogenesis shapes specialized transport tissues like xylem and phloem. Leaf morphogenesis steers leaf growth for photosynthesis, gas exchange, and transpiration. Floral morphogenesis orchestrates flower emergence and reproductive organs. Fruit and seed morphogenesis follows fertilization, culminating in fruit maturation and protective seed covering development. Secondary growth and lateral morphogenesis cause stem and root thickening, while tissue and organ regeneration facilitate damage recovery and even complete plant regeneration. Hormone-mediated morphogenesis, involving regulators like auxins and gibberellins, guides various growth aspects. These various types of morphogenesis collectively contribute to the overall growth, development, and adaptation of plants to their environments. The regulation of these processes is complex and involves intricate interactions between genetics, environmental cues, and hormonal signalling (Lichtenberger *et al.*, 2014).

Callus Morphogenesis

The term "callus" refers to the undifferentiated mass of cells that forms when plant tissues are cultured in vitro. Callus formation occurs as plant cells dedifferentiate from their specialized states, losing their specific characteristics and reverting to a more primitive, totipotent state (Sugimoto *et al.*, 2011). Callus morphogenesis refers to the process by which undifferentiated plant cells, known as callus cells, develop into organized structures resembling various parts of a plant, such as roots, shoots, and embryos. Skoog and Miller briefly described the "callus" as a disorganized developing tissue represented by a mass of dedifferentiated cells in their classic study on in vitro plant tissue culture. The further complete definition of a callus is as follows: a callus is an integrated system that develops as a result of the proliferation of surface or depth cells in various plant tissues. This system initially consists of homogeneous cells that gradually change into groups of heterogeneous cells with species-specific morphogenetic potential that are realized through various morphogenesis pathways (Kruglova *et al.*, 2018).

This concept is of great significance in plant tissue culture and regeneration techniques. Specific plant growth regulators or hormones to culture media could influence the process of callus morphogenesis. For example, the combination of auxins (such as indole-3-acetic acid, IAA) and cytokinins (such as kinetin) in the culture medium was found to induce callus formation. Once callus forms, its subsequent morphogenesis depends on the composition of the growth medium. Adjusting the concentrations of auxins and cytokinins in the medium can promote the development of specific structures within the callus, such as roots, shoots, and embryos. This ability to direct the differentiation and regeneration of callus cells is a critical aspect of plant tissue culture (Ramage and Williams, 2002).

Origin of Callus Morphogenesis

The origin of callus morphogenesis lies in the plant's remarkable ability to dedifferentiate and regenerate when exposed to specific conditions. The concept of callus morphogenesis can be traced back to the early 20th century when researchers observed that plant tissues have the capacity to regenerate when cultured in vitro (in a controlled environment). The earliest observations of tissue culture and regeneration date back to the late 19th and early 20th centuries (Thorpe, 2007). Researchers Gottlieb Haberlandt and Wilhelm Rauh experimented with isolating and culturing plant tissues on artificial media. Haberlandt's work on "totipotency" laid the foundation for the understanding of cells' ability to dedifferentiate and regenerate into whole plants. Origin of callus morphogenesis can be attributed to concept of totipotency and Pluripotency (Opatrný, 2013).

The Stages in Callus Morphogenesis

1. **Initiation:** Callus formation typically begins with the culture of plant explants (segments of plant tissues) on a nutrient-rich medium supplemented with plant growth regulators like auxins (*e.g.*, auxin-like compounds) and cytokinins. These hormones play a crucial role in inducing cell division and dedifferentiation, which are essential for callus formation.

2. **Dedifferentiation:** Once the explant is placed in culture, the plant cells at the cut edges start to lose their specialized characteristics and revert to a more primitive, totipotent state. This dedifferentiation process allows these cells to regain the ability to divide and differentiate into various cell types.

3. **Proliferation:** Under the influence of cytokinins, the dedifferentiated cells within the explant start to divide rapidly, forming a mass of undifferentiated cells known as a callus (white, friable, and irregularly shaped structure).

4. **Differentiation:** After the callus is formed, the next step involves the induction of specific differentiation pathways within the callus cells. This can be achieved by altering the balance of plant growth regulators in the culture medium. For instance, decreasing the concentration of cytokinins and increasing the concentration of auxins can promote the development of roots and shoots.

5. **Organogenesis**: As the callus differentiates, regions of the callus may begin to organize themselves into structures resembling various plant organs, such as roots, shoots, and embryos. These organized structures are called "organogenic calli."

6. **Maturation:** The differentiating structures within the callus continue to grow and develop, maturing into fully formed plant organs. Shoots may elongate, roots may develop root hairs, and embryogenic structures may become more distinct.

7. **Regeneration**: The matured organogenic calli, which have developed recognizable plant structures, can be induced to regenerate into whole plants. This usually involves transferring the developed structures onto a medium that supports further growth and development, often with reduced or modified concentrations of growth regulators.

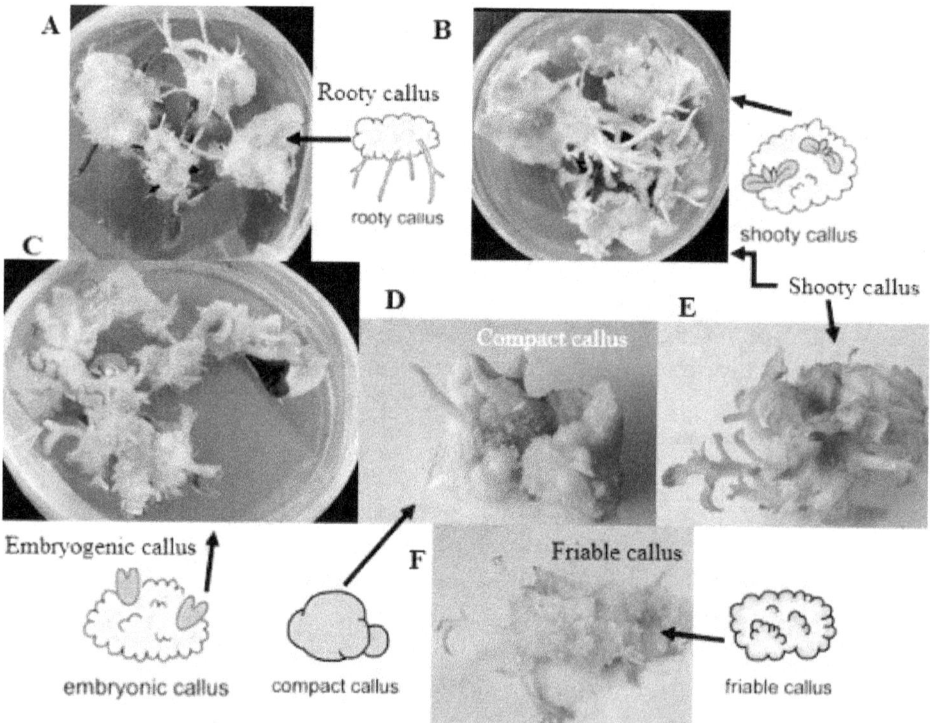

Figure 2.1: Morphology of Different Stages during Plant Organogenesis in *Nicotiana rustica*.

(A) Root regeneration from callus tissue. (B) Shooty callus. (C) Embryo regeneration from callus tissue. (D) Compact callus (E) Shoot regeneration from callus tissue (F) Shoot clumps regenerated from a friable callus (Bidabadi and Jain, 2020).

8. **Acclimatization:** Once the regenerated plants have grown sufficiently, they are transferred from the controlled laboratory conditions to more natural environments. This process, known as acclimatization, allows the plants to adapt to external conditions and thrive in their intended growth environment. The Figure 2.1 illustrates the morphology of different stages during plant organogenesis in *Nicotiana rustica* (Bidabadi and Jain, 2020)

Callus Morphogenesis Importance in Genome Editing

Callus morphogenesis plays a pivotal role in genome editing, particularly when employing techniques like CRISPR-Cas9, TALENs, and zinc finger nucleases. Genome editing involves precise modifications to an organism's DNA for desired trait development or genetic defect correction (Schaeffer and Nakata, 2015). Callus morphogenesis holds significance due to several reasons. Primarily, it facilitates efficient transformation as undifferentiated and rapidly dividing callus cells are more amenable to introducing editing machinery like CRISPR components compared to mature plant tissues. Additionally, the process of callus morphogenesis, involving cell dedifferentiation, enhances cell susceptibility to genome editing tools, thereby boosting editing efficiency. Embryogenic calli, resembling plant embryos within the callus, serve as ideal targets for genome editing, as they can develop into whole plants with desired traits (Efferth, 2019). Optimizing tissue culture conditions for callus induction and growth is essential for consistent and successful genome editing outcomes. Callus cultures are also valuable for generating genetic variation and haploidy, aiding in gene function study and trait improvement (Suprasanna *et al.*, 2014). After successful editing in callus cultures, the edited cells can be propagated and selected to develop into whole plants, amplifying the desired genetic changes. Multiplex editing, targeting multiple genome sites simultaneously, benefits from the undifferentiated nature of callus cells. Moreover, callus morphogenesis provides a solution for editing plant species that are hard to transform directly, serving as an intermediary step to overcome transformation challenges (Gordon-Kamm *et al.*, 2019).

In conclusion, callus morphogenesis holds critical importance in genome editing, particularly for efficient transformation, increased editing efficiency, embryogenic calli for trait development, tissue culture optimization, genetic variation generation, propagation and selection of edited cells, multiplex editing, and addressing challenges in editing recalcitrant plant species. Its significance in streamlining and enhancing various aspects of the genome editing process makes it an indispensable tool in advancing genetic modification and research in the plant sciences.

Transformation Techniques

The plant genetic transformation is a vital strategy for imparting desired genetic characteristics in the plant tissue for precise breeding. The technique being necessity for genome editing is bifurcated into indirect genetic transformation and direct genetic transformation. A vector (organism) is used to transform a target cell in the former case, whereas external forces delivers target genes into the target cells in the latter case. The different methods of gene transformation techniques were discussed briefly.

Indirect Transformation Technique

Indirect transformation technique or the vector mediated transformation is carried out using plasmids of gram-negative bacterium, *Agrobacterium* or with the help of plant viruses. The most commonly used vector for transformation is plasmid of *Agrobacterium* and the two species used include *Agrobacterium tumefaciens* and *Agrobacterium rhizogenes*, which are often referred to as the "natural genetic engineers". Apart from the bacterium species used for genetic engineering, an alternate vector which can be used for engineering the genome are the viral vectors. The viruses used are autonomously replicating either DNA viruses such as Bean Yellow Dwarf virus, Wheat Dwarf virus and Cabbage leaf curl virus or RNA viruses such as Tobacco leaf curl virus.

Agrobacterium Mediated Transformation

Agrobacterium mediated transformation is the most widely used method for genetic transformation, where there is stable expression of transgenes and also the occurrence of high number of introduced copies of the cassettes. Both the species have large plasmids, which can be modified and used as vectors. *Agrobacterium tumefaciens* has the Ti (Tumour inducing) plasmids which causes the root gall disease in dicot plants. In *Agrobacterium rhizogenes*, the Ri plasmid is present, which causes the hairy root disease in dicot plants. These plasmids have a size of 200kbp. These plasmids are capable of creating chimeric DNA in the host plant genome. The disease-causing regions of these plasmids are replaced with the gene of interest and used to cause transformation in the host plant genome. The commonly used plasmids are the Ti plasmids, since the Ti plasmids have the capability of infecting any tissues of the plant whereas the Ri plasmids infects the root tissues of the host plant.

The Ti plasmid (Figure 2.2) has the following components,

    a. T-DNA, with Right and Left Border which is transferred to the host. It has the genes for the production of plant hormones, auxin, cytokinin.

    b. *Vir* operons to recognize the plant signals and to transfer the T-DNA. *VirA* recognizes of the host signal in the form of monosaccharides, acetosyringone. It phosphorylates *VirG* and regulates other *vir* operons. *VirD* operon forms complex with T-DNA called T complex. Transfer of T-DNA is aided by the vir B protein encoded by *VirB* operon. *Vir E* (Vir E2 protein) operon aids in the protection of T complex. *VirE2* and *VirD2* imports the T complex into the nucleus, since it carries the nuclear localization signals.

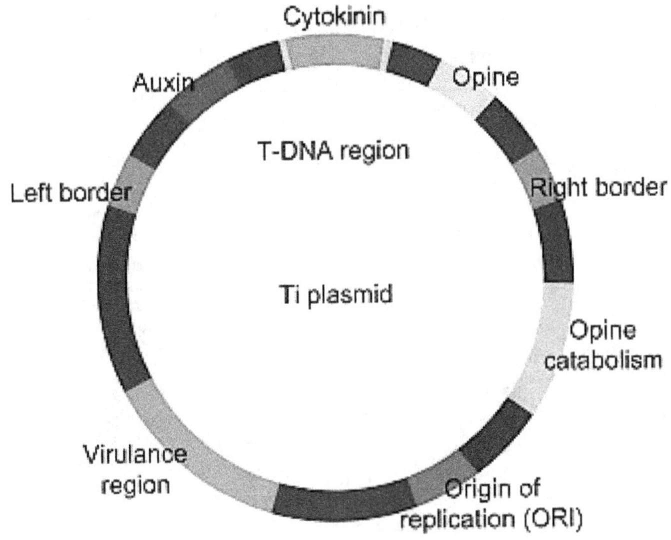

**Figure 2.2: Ti Plasmid
(Tzotzos *et al.*, 2009).**

VirE2 INTEGRATING PROETINS (VIP), plant proteins which helps in the localization of T complex in the nucleus. VIP1 protein integrates with *Vir E2* and VIP2 for T-DNA integration in plants.

c. *rep* region, for replication of the plasmid. *repA, repB and repC* help in the replication and maintenance of copy number (Tabata *et al.*, 1989).

d. *tra and trb* for direct conjugal transfer of the plasmid. The series of steps from signal induction to T-DNA integration were depicted in Figure 2.3.

Ti plasmids used for gene integration are modified such that the Tumour inducing genes are replaced with the gene of interest and closely linked selectable marker genes such as the antibiotic or herbicide resistance. T-DNA genes are more likely to be expressed in host plant cells because of their lower G + C composition (Suzuki *et al.*, 2009).

To improve the efficiency and to avoid the insertion of the T-DNA regions other than the gene of interest, alternate vector systems are used. It includes the disarmed Ti plasmid, binary vector system *etc.*, *Agrobacterium* strains having high transformation efficiency are used or their genome is modified by either addition or mutation, to increase the efficiency of the transformation. Ti plasmids as such cannot be used because of its very large size and its low copy number in *Agrobacterium* and the oncogenes, which cause the development of abnormal phenotypes. These shortcomings are overcome using the development of T-DNA based vectors, namely co- integrate vectors and binary vectors. These vectors make use of the Ti plasmid, which lack the oncogenes responsible for the production of plant hormones called disarmed Ti plasmid.

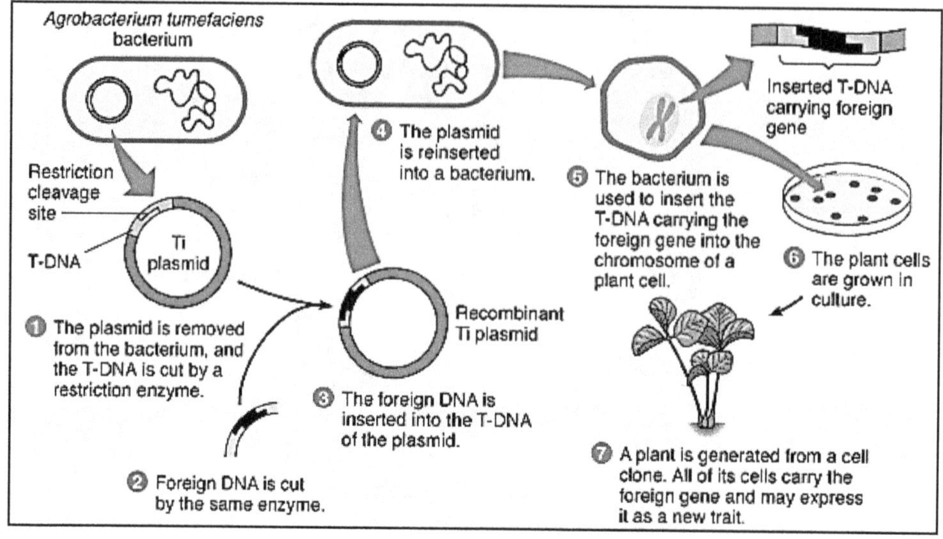

Figure 2.3: Protocol for *Agrobacterium* T-DNA Integration into the Host Plant (https://sphweb.bumc.bu.edu/otlt/MPH-Modules/PH/GMOs/GMOs3.html).

Viral Vector Mediated Transformation

The viral vectors have been used for the production of useful proteins. Its genome organization and machinery, makes it widely useful for altering the eukaryotic genome. The major disadvantage of host specificity of *Agrobacterium* mediated transformation can be culled out using viral vectors (Zaidi *et al.*, 2017). For monocots transformation, RNA viruses such as wheat streak mosaic virus and Barley stripe mosaic virus are widely used (Lee *et al.*, 2012). DNA viruses belonging to the family Geminiviridae which infect a wide range of host plant species from monocots like wheat to dicots including legumes, tomato, cotton *etc.*, are used widely for transformation (Rey *et al.*, 2012). Other markable features include, it has a small genome (appx. 2.8 kb) it requires only one protein RecA for replication, uses natural promoter within the intergenic region for expression (Baltes *et al.*, 2014), has high replication efficiency and enhanced target efficiency (Hanley- Bowdain *et al.*, 2013). However, the ability of the virus for transport into the host is restricted which can be overcome using the non-infectious replicons just by replacing the genes responsible for infection. The movement protein, coat protein and the coding sequences of the virus is removed, which thereby eliminates the cell-to-cell movement and its transmission through the vector insects and increases the copy number per cell. Recent studies indicate the use of viral vectors for the delivery of genome editing into the target host. Gene targeting was appx. 12 times higher in BeYDV than that of using *Agrobacterium* mediated T-DNA delivery (Cermark *et al.,* 2015). Wang *et al.* (2017) showed that the CRISPR/Cas mediated gene targeting and the HDR (19.4 per cent high) was highly efficient in rice when Wheat Dwarf Virus based vector was used. These vectors had led to the development of virus-based gRNA delivery system, where the viral vectors cause the transient delivery of sgRNA for targeting the gene of interest and the Cas9 is overexpressed, which is an alternate for Virus Induced Gene Silencing (VIGS).

Single stranded RNA virus, Tobacco rattle virus (TRV), belonging to Vigaviridae family is transmitted through nematode and through seeds. It has many advantages such as, it can be used in wide range of plant species, small genome size of the virus, viral genome does not integrate into plant genomes. TRV has two genomes, TRV1 and TRV2. TRV1 was used for viral movement and has genes for replicase protein. TRV2 genome varies for different isolates and has genes for coat protein and non- structural proteins. These non- structural proteins are replaced with multiple cloning sites to replace with fragments of interest (Senthil Kumar and Mysore, 2014). Marton *et al.* (2010) used this vector for delivery of ZFN. Ali *et al.* (2015a) used TRV mediated sgRNA delivery in *Nicotiana bentamiana* transgenic lines.

Direct Transformation Technique

The direct gene transfer method relies on delivery of large amount of naked DNA. It has wider application in cereal crop transformation. Various direct gene transfer method were followed in different crops, inspite of its complex integration patterns and higher copy number.

Particle Bombardment Technique

Particle bombardment technique also known as particle gun, the gene gun, the bio blaster and microprojectile bombardment method was first developed and described by John Sanford in onion tissues in epidermal tissues with transgene chloramphenicol acetyl transferase (CAT) (Sanford *et al.,* 2000). This technique can be used in broad range of plant species even where vector-based transformation is not possible and one of the physical methods for the direct introduction of DNA into the plant genome. The easiness and the target material's diversity seeks attention of this method, whereas high cost, reduced transformation rate and insecure exogenous DNA were the lacuna (Keshavareddy *et al.,* 2018). Smaller DNA fragments of less than 10kb are preferred owing to the breakage of larger fragments or weak attachment to metals resulting in disordered DNA integration. This method have been employed for targeted insertions in rice and soybean (Begemann *et al.,* 2017; Bonawitz *et al.,* 2019).

Procedure

DNA particles are coated on microcarriers called microprojectile like gold, tungsten, platinum, palladium, rhodium, iridium but gold and tungsten are commonly used.

↓

These microprojectiles are delivered with high velocity using gene gun. Initially gun powder driven gun was used which was replaced by helium driven guns

↓

The microprojectile is loaded onto a macrocarrier usually aqueous slurry loaded on the end of a small plastic bullet

↓

This microprojectile with macrocarrier is loaded onto the gun and fired with high velocity and a stopping screen catches the macrocarrier allowing only microprojectiles to proceed further

↓

Then the microprojectile containing the DNA reaches the target tissue in partial vacuum conditions and the DNA is delivered into the cells of the tissue.

Modified Bombardment Devices

Other types of devices other than gun powder device is used to increase the efficiency which are as follows: Helium modified bombardment device, Accel Particle gun, Particle inflow gun, Micro-targeting device, Helios gene gun.

Electroporation

An electric pulse of high field strength is used to create a temporary pore in the cell membrane which aids in DNA uptake. This is done in a chamber of electroporator. Using linear/circular DNA with field strength of 1.25 kV/ cm along with PEG can improve the protoplast transforming efficiency. There is a theory which states that application of electric field activates the enzyme permease which is responsible for formation of pores on the cell membrane. In comparison with *Agrobacterium* and gene gun method, the benefits of this method includes low cost, rapid application and high stable transformation rate (Kar *et al.,* 2018). The major drawback is that the fierceful electric pulses destroys the naked DNA and works well with very few species (Yan *et al.,* 2022).

Chemical (PEG) Method

DNA of interest along with selectable marker in the ratio of 3:1 to 10:1 ratio is taken in plasmid form (linear double stranded one is preferred than supercoiled one) and is added to a protoplast suspension. To this 40 per cent PEG 4000 + Mannitol + Calcium Nitrate is added slowly. This PEG aids in transformation of the protoplasts and the target DNA is integrated randomly into the genome. Transformed protoplasts are then taken and cultured accordingly. This method aids in both transient and stable transformation.

Micro-injection

Recipient cells are fixed on a poly-lysine treated glass surface, holded with a suction holding pipette and they are injected with a micro needle containing DNA to be transferred. These protoplasts/cells can be cultured. Sometimes this method results in chimera transformation. It has been widely followed in animal cells as the syringe has a difficulty in penetrating into the cell wall and injecting exogenous DNA into the thick cell walls of plants (Lörz *et al.*, 1981).

Liposome

The exogenous DNA is introduced into the protoplasts by protoplast endocytosis or plasma membrane fusion. DNA-lipid complex forms as a result of mixing of liposomes and DNA, which is consequently mixed with protoplast suspension (complemented with PEG) (Gad *et al.*, 1990). The positive liposome and negative DNA were attracted towards each other with a release of plasmids into the target cells (van Wordragen *et al.*, 1997).

Sonication

This method uses ultrasound pulses of selected intensity and duration and transforms protoplasts, cells and small tissues are prepared as a suspension. Plasmid DNA is added in the microfuge tube in a sonication medium which already has cell suspension in it. These cells are then cultured and grown.

Laser Micro Beam

DNA enters into the cell through a self-healing hole made with a laser microbeam. This method is even applicable for cells with hard walls and tissues. Thus, this method can be effectively utilized for plants where protoplast regeneration is difficult.

Silicon Carbide Fiber-vortex (Silicone Whiskers)

A mixture of plasmid DNA with selectable marker gene, silicon carbide fibres and explant tissue are taken in an eppendorf tube and vortexed. These silicone carbide fibers are tiny micro-needle like elements with 0.5 µm diameter and 10-80 µm in length, which is thought to penetrate cell wall and deliver the DNA which is coated onto it. This method is simple and efficient without the need for any special equipments. However, carcinogenic nature of silicon carbide fibres and the damage encountered during operation hinders the regeneration efficiency, resulting in low conversion efficiency (Saeed *et al.*, 2016).

Aerosol Beam Injection

DNA is delivered *via* a simple aerosol suspension. RNA and protein can also be inserted using this method. Aerosol droplets mixed with required DNA particles is subjected to a pressure differential such that these droplets are accelerated to a supersonic speed enabling them to penetrate cells and deliver the DNA.

Pollen-Tube Pathway Mediated Transformation

This method involves removal of stigma soon after pollination from recipient parent and addition of exogenous DNA solution drop by drop to the severed style (Luo *et al.*, 1989). As the pollen tube grows, the DNA gets transported along the tube and gets integrated with the fertilized egg, which is still undivided at its embryogenic stage and hence seen in the transformed seed. The method evades poor regeneration and genotype limitation and is not widely used due to the limitation of natural flowering period (Su *et al.*, 2023). The direct gene transfer methods employed in different crop species is listed below in Table 2.1.

Table 2.1: Direct Transformation Method Employed in different Crop Plants

Crop Species	Explant Used	Genes	Gene Delivery Method	Efficiency (Per cent)	References
Rice	Calli	*Cpf1* and crRNA	Particle bombardment	8 per cent	Begemann *et al.*, 2017
Sorghum	Immature embryos	*NptII*	Particle bombardment	46.6 per cent	Belide *et al.*, 2017
Maize	Calli	*VHb*	Particle bombardment	-	Du *et al.*, 2016
Barley	Seeds	*OsWRKY70*, *gus* and *OsWRKY53*	Particle bombardment	-	Zhang *et al.*, 2015
Blackgram	Embryonic axis	*ChiB*	Particle bombardment	13 per cent	Das *et al.*, 2018
Soybean	Embryonic axis	*Hpt*, HDRdonor and ZFN expression constructs	Particle bombardment	About 2.84 per cent	Bonawitz *et al.*, 2019
Wheat	Immature embryos	*bar* and *uid*A	Electroporation	0.4 per cent	Sorokin *et al.*, 2000
Tomato	Leaves	Fe and Mg	Liposome	33 per cent	Karny *et al.*, 2018

Crop Species	Explant Used	Genes	Gene Delivery Method	Efficiency (Per cent)	References
Cotton	Embryogenic callus	*Gus* and *AVP1*	Silicon carbide whisker	Upto 94 per cent	Asad *et al.*, 2008
Peanut	Callus	chitinase and*hygromcin*	Silicon carbide whisker	6.88 per cent	Akram *et al.*, 2016
Barley	Protoplasts	*Act1, gus* and *nos*	Microinjection	-	Holm *et al.*, 2000
Oilpalm	Protoplasts	*GFP*	Microinjection	10-7.6 per cent	Masani *et al.*, 2014
Maize	Pollen	*GFP*	Pollen tube pathay	0.86 per cent	Yang *et al.*, 2009
Cotton	Pollen	*nptll*	Pollen tube pathay	-	Wang *et al.*, 2019
Melon	Pollen	*Fom-2*	Pollen tube pathay	3.28 per cent and 4.26 per cent	Li *et al.*, 2019
Peanut	Pollen	*AhBl-1*	Pollen tube pathay	50 per cent	Zhou *et al.*, 2023

New Transformation Techniques

There exist some drawbacks of conventional transformation techniques like *Agrobacterium* mediated gene transfer and Particle bombardment method such as cell damage, low transformation efficiency and integration at random sites. To overcome these limitations, scientists have developed new methods of transformation such as nanoparticle mediated delivery and transformation through CRISPR-CAS and ZFN. The latter helps in avoiding insertion of new gene sequence at random sites thus helping in preventing normal processing of housekeeping genes.

Nanoparticle Mediated Delivery

This method employs magnetic nanoparticles, peptide nanoparticles, layered double hydroxide nanosheets (LDH), DNA nanostructures, and carbon nanotubes to deliver DNA into intact plant cells directly without any external force. The different methods of nanoparticle mediated delivery method is illustrated in Figure 2.4.

a) Magnetic-Nanoparticle-Mediated Gene Delivery

The DNA is first wrapped upon a magnetic nanoparticle (MNP) which is then introduced into pollen via magnetofection. This MNP-DNA complex enters through pollination and gets integrated into the offspring genome which gives us transgenic plants in the next generation (Zhao *et al.*, 2017). This method increases efficiency of transformation, independent of species, do not rely on tissue culture techniques, requires less time and is amenable for co-transformation of multigene all of which are of great importance to hasten the process of producing transformed plants.

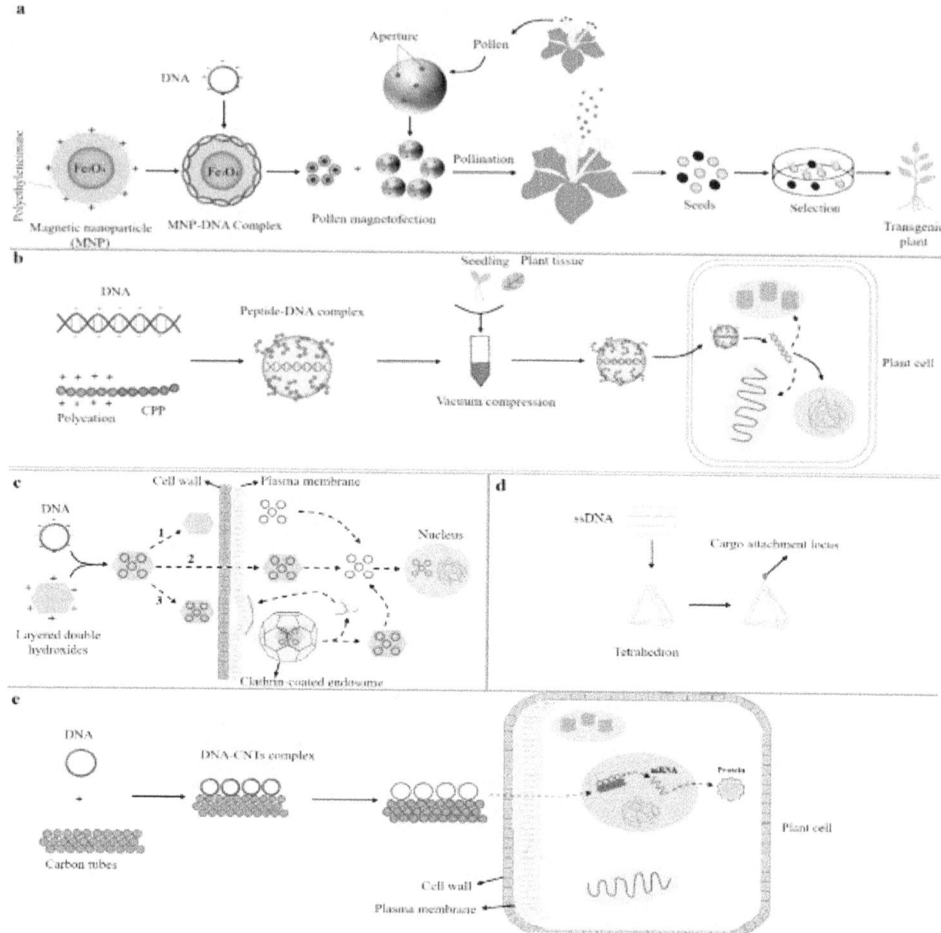

Figure 2.4: Different Methods of Nanoparticle Mediated Delivery (Su *et al.*, 2023).

b) *Peptide-Mediated Gene Delivery*

This method uses a kind of low molecular weight peptides called cell-penetrating peptides (CPPs) which aids in cell absorption of materials like nanoparticles, small compounds and a stretch of DNA. Negatively charged DNA is first bound with these CPPs and this peptide-DNA complex then entered into cell through compression or vacuum. Chang *et al.* (2005), revealed that CPPs could transfer protein to different tissues of tomato (dicots) and onion (monocots), demonstrating that CPPs could deliver exogenous biomolecules to entire plant cells via the cell wall and lipid membrane. According to recent research, organelle-targeting peptides could deliver DNA to specific organelles in intact plants, including nucleus, mitochondria, and chloroplasts.

c) *Layered-Double-Hydroxide-Mediated Gene Delivery*

Layered double hydroxides (LDHs) are a type of ionic layered compound that has positively charged sublayers which contains charge-compensating anions, solvates and an intermediate layer consisting of charge-balancing anions and co-embedded water (Xu *et al.*, 2006). This cationic nature of LDHs enables them to strongly bind to negatively charged DNA. LDHs delivers DNA through three different pathways: first involves transport of DNA/RNA to cytoplasm through plasma membrane leaving behind LDH complex, second involves transportation of both LDH-DNA/RNA complex into cytoplasm through non-intracellular pathway whereas the third involves internalization of LDH-DNA/RNA complex into plant cells through endocytosis.

d) *DNA-Nanostructure-Mediated Gene Delivery*

This technology employs precise base pairing approach where artificial single stranded DNA Sequences is assembled into nanostructures of predefined size, shapes and geometrics upon which our target DNA, RNA and protein gets attached at the cargo attachment site (Sharma *et al.*, 2018). A GFP gene was silenced in transgenic tobacco (mGFP5) by using this method employing 21bp long siRNAs which prevented the constitutive expression of that gene in the nucleus. A comparative study between free siRNA and DNA nanostructure loaded siRNA revealed that degradation was negligible in the latter which resulted in complete expression of the trait.

e) *Carbon-Nanotube-Mediated Gene Delivery*

There are two types of carbon nanotubes single walled carbon nanotubes (SWCNTs) and multi walled carbon nanotubes (MWCNTs). SWCNTs are cylindrical structures made up of graphene layers measuring about 0.7-3 mm in diameter whereas MWCNTs are made up of multiple single walled carbon nanotubes measuring 220 mm in diameter. A DNA-CNT complex is formed which is capable of penetrating the cell wall and entering the cytoplasm. Several success stories have been reported in various crops like tobacco, wheat and cotton by employing polyethylene imine modified carbon nanotubes (PEI-CNTs) which are capable of binding DNA electrostatically and also it prevents from nuclease degradation.

CRISPR-Cas/ZFN Mediated Targeted DNA Insertion

In all the above said methods, the target DNA will be inserted randomly in the genome which may cause insertion in housekeeping genes and prevents its transcription which is essential for normal plant growth and development.

In order to avoid this, target DNA can be inserted at specific sites in genome without affecting native genes by CRISPR-Cas system and zinc finger nuclease (ZFN) system can be used which guides the target DNA to a specified site in the genome. After inducing double strand break at the target site, the DNA which is delivered along with the Crispr Cas or ZFN system is used as a template in the homology repair mediated pathway to produce the insertion at the specific site.

Future Perspectives on Gene Transformation and Callus Morphogenesis

Gene transformation and callus morphogenesis are two dynamic aspects that hold immense promise for shaping the future of biology, biotechnology, and agriculture (Efferth, 2019). These areas are driven by groundbreaking technologies and innovative applications that are poised to revolutionize various aspects of human life. As we peer into the horizon of possibilities, it becomes clear that these advancements will have far-reaching impacts on precision editing, synthetic biology, agriculture, and ethical considerations.

Precision Genome Editing and Epigenetic Editing

One of the most remarkable advancements in gene transformation is precision genome editing. While CRISPR/Cas9 technology has already transformed the landscape of genetic modification, the future lies in refining its precision and minimizing off-target effects (Shah *et al.*, 2018). Prime editing, a recent development, exemplifies this progress by enabling accurate single-nucleotide changes. Such advancements hold great potential for addressing genetic disorders with pinpoint accuracy, ushering in a new era of safer and more effective gene therapies (Joynt, 2022). Furthermore, the emergence of epigenetic editing offers a unique perspective on gene transformation. Scientists are exploring ways to modify epigenetic marks, thereby influencing gene expression without altering the underlying DNA sequence. This opens avenues for therapeutic interventions in conditions with epigenetic components, such as cancer. The fusion of DNA-modifying enzymes with programmable DNA-binding domains presents an exciting approach to reshape cellular identity and behavior (Chakraborty *et al.*, 2016).

Synthetic Biology and Multi-Gene Engineering

The concept of designing synthetic organisms with tailored functions is no longer confined to science fiction. Recent achievements, such as the construction of a smaller yet functional synthetic yeast chromosome, showcase the potential of synthetic biology. This innovation has implications for bioproduction, environmental remediation, and even the creation of organisms optimized for specific tasks. Moreover, the engineering of multi-gene cassettes, which confer complex traits in crops, highlights a transformative approach to addressing global challenges like food security and sustainability (Que *et al.*, 2010).

Beyond Model Organisms and Ethical Considerations

The scope of gene transformation techniques extends beyond model organisms and human applications. Manipulating microbiomes and engineering non-model organisms like gut bacteria for therapeutic purposes demonstrates the potential to revolutionize healthcare and environmental solutions. However, this rapid progress raises important ethical and regulatory considerations (Russell *et al.*, 2022). The ongoing dialogue surrounding human germline editing emphasizes the importance of establishing clear ethical guidelines. Regulatory bodies worldwide face the delicate task of fostering innovation while ensuring responsible and accountable use of gene transformation technologies (Ormond *et al.*, 2017).

As we look ahead, the integration of callus morphogenesis with gene transformation presents a novel dimension. Advances in omics technologies, epigenetic regulation, single-cell analysis, and 3D tissue culture techniques promise to unravel the intricacies of cellular reprogramming and differentiation (Akkerman *et al.*, 2017). Synthetic biology applications offer the potential to engineer efficient regeneration processes, impacting fields ranging from medicine to agriculture. Climate resilience and crop improvement stand to benefit from leveraging callus morphogenesis to develop stress-tolerant crops. In conclusion, the future of gene transformation and callus morphogenesis holds a tapestry of opportunities. Through scientific exploration, technological innovation, and ethical considerations, these techniques will continue to shape our understanding of biology and redefine the boundaries of possibility. As these advancements unfold, it is imperative to ensure responsible use and equitable distribution of benefits to harness the full potential of these transformative technologies.

REFERENCES

Akkerman, N., and Defize, L. H. (2017). Dawn of the organoid era: 3D tissue and organ cultures revolutionize the study of development, disease, and regeneration. *Bioessays*, *39*(4), 1600244.

Akram, Z., Ali, S., Ali, G. M., Zafar, Y., Shah, Z. H., and Alghabari, F. (2016). Whisker-mediated transformation of peanut with chitinase gene enhances resistance to leaf spot disease. *Crop Breeding and Applied Biotechnology*, 16, 108-114.

Ali, Z., Abul-Faraj, A., Li, L., Ghosh, N., Piatek, M., Mahjoub, A.,. and Mahfouz, M. M. (2015). Efficient virus-mediated genome editing in plants using the CRISPR/Cas9 system. *Molecular plant*, 8(8), 1288-1291.

Arber, W. (2014). Horizontal gene transfer among bacteria and its role in biological evolution. *Life*, 4(2), 217-224.

Asad, S., Mukhtar, Z., Nazir, F., Hashmi, J. A., Mansoor, S., Zafar, Y., and Arshad, M. (2008). Silicon carbide whisker-mediated embryogenic callus transformation of cotton (*Gossypium hirsutum* L.) and regeneration of salt tolerant plants. *Molecular biotechnology*, 40, 161-169.

Baltes, N. J., Gil-Humanes, J., Cermak, T., Atkins, P. A., and Voytas, D. F. (2014). DNA replicons for plant genome engineering. *The Plant Cell*, 26(1), 151-163.

Begemann, M. B., Gray, B. N., January, E., Gordon, G. C., He, Y., Liu, H.,. and Oufattole, M. (2017). Precise insertion and guided editing of higher plant genomes using *Cpf1* CRISPR nucleases. *Scientific reports*, 7(1), 11606.

Belide, S., Vanhercke, T., Petrie, J. R., and Singh, S. P. (2017). Robust genetic transformation of sorghum (Sorghum bicolor L.) using differentiating embryogenic callus induced from immature embryos. *Plant methods*, 13(1), 1-12.

Bidabadi, S. S., and Jain, S. M. (2020). Cellular, molecular, and physiological aspects of in vitro plant regeneration. *Plants*, 9(6), 702.

Bonawitz, N. D., Ainley, W. M., Itaya, A., Chennareddy, S. R., Cicak, T., Effinger, K.,. and Pareddy, D. R. (2019). Zinc finger nucleasemediated targeting of multiple transgenes to an endogenous soybean genomic locus via nonhomologous end joining. *Plant biotechnology journal*, 17(4), 750-761.

Èermák, T., Baltes, N. J., Èegan, R., Zhang, Y., and Voytas, D. F. (2015). High-frequency, precise modification of the tomato genome. *Genome biology*, 16, 1-15.

Chakraborty, K., Veetil, A. T., Jaffrey, S. R., and Krishnan, Y. (2016). Nucleic acid–based nanodevices in biological imaging. *Annual Review of Biochemistry*, 85, 349-373.

Das, D. K. (2018). Expression of a bacterial chitinase (*ChiB*) gene enhances resistance against Erysiphae polygoni induced powdery mildew disease in the transgenic black gram (*Vigna mungo* L.)(cv. T9). *American Journal of Plant Sciences*, 9(8), 1759-1770.

de Almeida, M., Graner, E.M., Brondani, G.E., de Oliveira, L.S., Artioli, F.A., de Almeida, L.V., Leone, G.F., Baccarin, F.J.B., de Oliveira, A.P., and Cordeiro, G.M.; *et al.* (2015). Plant morphogenesis: Theorical bases. *Adv. Forest. Sci.*, 2, 13–22.

Du, H., Shen, X., Huang, Y., Huang, M., and Zhang, Z. (2016). Overexpression of Vitreoscilla hemoglobin increases waterlogging tolerance in *Arabidopsis* and maize. *BMC Plant Biology*, 16(1), 1-11.

Efferth, T. (2019). Biotechnology applications of plant callus cultures. *Engineering*, 5(1), 50-59.

Gaarslev, N., Swinnen, G., and Soyk, S (2021). Meristem transitions and plant architecture–Learning from domestication for crop breeding. *Plant Physiol.*, 187, 1045–1056.

Gad, A. E., Rosenberg, N., and Altman, A. (1990). Liposomemediated gene delivery into plant cells. *Physiologia Plantarum*, 79(1), 177-183.

Gbashi, S., Adebo, O., Adebiyi, J. A., Targuma, S., Tebele, S., Areo, O. M.,. and Njobeh, P. (2021). Food safety, food security and genetically modified organisms in Africa: a current perspective. *Biotechnology and Genetic Engineering Reviews*, 37(1), 30-63.

Gordon-Kamm, B., Sardesai, N., Arling, M., Lowe, K., Hoerster, G., Betts, S., and Jones, T. (2019). Using morphogenic genes to improve recovery and regeneration of transgenic plants. *Plants*, 8(2), 38.

Guo, K., Huang, C., Miao, Y., Cosgrove, D. J., and Hsia, K. J. (2022). Leaf morphogenesis: The multifaceted roles of mechanics. *Molecular plant.*

Hanley-Bowdoin, L., Bejarano, E. R., Robertson, D., and Mansoor, S. (2013). Geminiviruses: masters at redirecting and reprogramming plant processes. *Nature Reviews Microbiology*, 11(11), 777-788.

Holm, P. B., Olsen, O., Schnorf, M., Brinch-Pedersen, H., and Knudsen, S. (2000). Transformation of barley by microinjection into isolated zygote protoplasts. *Transgenic Research*, 9, 21-32.

Joynt, A. T. (2022). *Investigation of precision therapies for splice variants* (Doctoral dissertation, Johns Hopkins University).

Kar, S., Loganathan, M., Dey, K., Shinde, P., Chang, H. Y., Nagai, M., and Santra, T. S. (2018). Single-cell electroporation: current trends, applications and future prospects. *Journal of Micromechanics and Microengineering*, 28(12), 123002.

Karny, A., Zinger, A., Kajal, A., Shainsky-Roitman, J., and Schroeder, A. (2018). Therapeutic nanoparticles penetrate leaves and deliver nutrients to agricultural crops. *Scientific reports*, 8(1), 7589.

Kelley, N. J., Whelan, D. J., Kerr, E., Apel, A., Beliveau, R., and Scanlon, R. (2014). Engineering biology to address global problems: Synthetic biology markets, needs, and applications. *Industrial Biotechnology*, 10(3), 140-149.

Keshavareddy, G., Kumar, A. R. V., and Ramu, V. S. (2018). Methods of plant transformation-a review. Int. J. Curr. *Microbiol. Appl. Sci*, 7(7), 2656-2668.

Kruglova, N. N., Titova, G. E., and Seldimirova, O. A. (2018). Callusogenesis as an in vitro morphogenesis pathway in cereals. *Russian Journal of Developmental Biology*, 49, 245-259.

Lee, W. S., Hammond-Kosack, K. E., and Kanyuka, K. (2012). Barley stripe mosaic virus-mediated tools for investigating gene function in cereal plants and their pathogens: virus-induced gene silencing, host-mediated gene silencing, and virus-mediated overexpression of heterologous protein. *Plant Physiology*, 160(2), 582-590.

Li, X., Liu, L., Liu, X., Hou, Y., Xu, B., Luan, F., and Wang, X. (2019). Construction of expression vectors of the melon resistance gene fom-2 and genetic transformation. *Pak. J. Bot*, 51(1), 125-132.

Lichtenberger, R., FurutaMiyashima, K., Hellmann, E., and Helariutta, Y. (2014). Phloem cell development. *Plant Cell Wall Patterning and Cell Shape*, 379-399.

Lörz, H., Paszkowski, J., DierksVentling, C., and Potrykus, I. (1981). Isolation and characterization of cytoplasts and miniprotoplasts derived from protoplasts of cultured cells. *Physiologia Plantarum*, 53(3), 385-391.

Luo, Z. X., and Wu, R. (1989). A simple method for the transformation of rice via the pollen-tube pathway. *Plant Molecular Biology Reporter*, 7, 69-77.

Mahfouz, M. M., Piatek, A., and Stewart Jr, C. N. (2014). Genome engineering via TALENs and CRISPR/Cas9 systems: challenges and perspectives. *Plant biotechnology journal*, 12(8), 1006-1014.

Martin-Ortigosa, S., Trewyn, B. G., and Wang, K. (2017). Nanoparticle-mediated recombinase delivery into maize. *Site-Specific Recombinases: Methods and Protocols*, 169-180.

Marton, I., Zuker, A., Shklarman, E., Zeevi, V., Tovkach, A., Roffe, S.,. and Vainstein, A. (2010). Nontransgenic genome modification in plant cells. *Plant physiology*, 154(3), 1079-1087.

Masani, M. Y. A., Noll, G. A., Parveez, G. K. A., Sambanthamurthi, R., and Prüfer, D. (2014). Efficient transformation of oil palm protoplasts by PEG-mediated transfection and DNA microinjection. *PloS one*, 9(5), e96831.

Ochman, H., Lawrence, J. G., and Groisman, E. A. (2000). Lateral gene transfer and the nature of bacterial innovation. *nature*, 405(6784), 299-304.

Opatrný, Z. (2013). From Nìmec and Haberlandt to plant molecular biology. In *Applied Plant Cell Biology: Cellular Tools and Approaches for Plant Biotechnology* (pp. 1-36). Berlin, Heidelberg: Springer Berlin Heidelberg.

Ormond, K. E., Mortlock, D. P., Scholes, D. T., Bombard, Y., Brody, L. C., Faucett, W. A.,. and Young, C. E. (2017). Human germline genome editing. *The American Journal of Human Genetics*, 101(2), 167-176.

Que, Q., Chilton, M. D. M., de Fontes, C. M., He, C., Nuccio, M., Zhu, T.,. and Shi, L. (2010). Trait stacking in transgenic crops: challenges and opportunities. *GM crops*, 1(4), 220-229.

Ramage, C. M., and Williams, R. R. (2002). Mineral nutrition and plant morphogenesis. *In Vitro Cellular and Developmental Biology-Plant*, 38, 116-124.

Rey, M. E., Ndunguru, J., Berrie, L. C., Paximadis, M., Berry, S., Cossa, N.,. and Esterhuizen, L. L. (2012). Diversity of dicotyledenous-infecting geminiviruses and their associated DNA molecules in Southern Africa, including the South-West Indian Ocean Islands. Viruses, 4(9), 1753-1791.

Russell, B. J., Brown, S. D., Siguenza, N., Mai, I., Saran, A. R., Lingaraju, A.,. and Zarrinpar, A. (2022). Intestinal transgene delivery with native E. coli chassis allows persistent physiological changes. *Cell*, *185*(17), 3263-3277.

Saeed, T., and Shahzad, A. (2016). Basic principles behind genetic transformation in plants. Biotechnological strategies for the conservation of medicinal and ornamental climbers, 327-350.

Sanford, J. C. (2000). The development of the biolistic process. In Vitro Cellular and Developmental Biology-*Plant*, 36(5), 303-308.

Sattler, Rolf. "Process morphology: structural dynamics in development and evolution." *Canadian Journal of Botany* 70.4 (1992): 708-714.

Schaeffer, S. M., and Nakata, P. A. (2015). CRISPR/Cas9-mediated genome editing and gene replacement in plants: transitioning from lab to field. *Plant Science*, *240*, 130-142.

Senthil-Kumar, M., and Mysore, K. S. (2014). Tobacco rattle virus–based virus-induced gene silencing in *Nicotiana benthamiana*. *Nature protocols*, 9(7), 1549-1562.

Shah, S. Z., Rehman, A., Nasir, H., Asif, A., Tufail, B., Usama, M., and Jabbar, B. (2018). Advances in research on genome editing CRISPR-Cas9 technology. *Journal of Ayub Medical College Abbottabad*, *31*(1), 108-122.

Sharma, A., Vaghasiya, K., Verma, R. K., and Yadav, A. B. (2018). DNA nanostructures: chemistry, self-assembly, and applications. In *Emerging Applications of Nanoparticles and Architecture Nanostructures* (pp. 71-94). Elsevier.

Singh, R. (2011). Facts, growth, and opportunities in industrial biotechnology. *Organic Process Research and Development*, *15*(1), 175-179.

Sorokin, A. P., Ke, X. Y., Chen, D. F., and Elliott, M. C. (2000). Production of fertile transgenic wheat plants via tissue electroporation. Plant Science, 156(2), 227-233.

Su, W., Xu, M., Radani, Y., and Yang, L. (2023). Technological Development and Application of Plant Genetic Transformation. *International Journal of Molecular Sciences*, 24(13), 10646.

Sugimoto, K., Gordon, S. P., and Meyerowitz, E. M. (2011). Regeneration in plants and animals: dedifferentiation, transdifferentiation, or just differentiation?. *Trends in cell biology*, 21(4), 212-218.

Suprasanna, P., Mirajkar, S. J., Patade, V. Y., and Jain, S. M. (2014). Induced mutagenesis for improving plant abiotic stress tolerance. In *Mutagenesis: exploring genetic diversity of crops* (p. e30765). Wageningen Academic Publishers.

Suzuki, K., Tanaka, K., Yamamoto, S., Kiyokawa, K., Moriguchi, K., and Yoshida, K. (2009). Ti and Ri plasmids. *Microbial megaplasmids*, 133-147.

Tabata, S. A. T. O. S. H. I., Hooykaas, P. J., and Oka, A. T. S. U. H. I. R. O. (1989). Sequence determination and characterization of the replicator region in the tumor-inducing plasmid pTiB6S3. *Journal of bacteriology, 171*(3), 1665-1672.

Thorpe, T. A. (2007). History of plant tissue culture. *Molecular biotechnology, 37*, 169-180.

Tzotzos, G. T., Head, G. P., and Hull, R. (2009). Setting the context: agriculture and genetic modification. *Genetically Modified Plants*, 1-32.

van Wordragen, M., Shakya, R., Verkerk, R., Peytavis, R., van Kammen, A., and Zabel, P. (1997). Liposome-mediated transfer of YAC DNA to tobacco cells. *Plant Molecular Biology Reporter*, 15, 170-178.

Wan, Q., Zhai, N., Xie, D., Liu, W., Xu, L. and WOX11 (2023): The founder of plant organ regeneration. *Cell Regen.* 12, 1.

Wang, M., Lu, Y., Botella, J. R., Mao, Y., Hua, K., and Zhu, J. K. (2017). Gene targeting by homology-directed repair in rice using a geminivirus-based CRISPR/Cas9 system. *Molecular plant*, 10(7), 1007-1010.

Wang, M., Sun, R., Zhang, B., and Wang, Q. (2019). Pollen tube pathway-mediated cotton transformation. Transgenic Cotton: Methods and Protocols, 67-73.

Xu, Z. P., Zeng, Q. H., Lu, G. Q., and Yu, A. B. (2006). Inorganic nanoparticles as carriers for efficient cellular delivery. *Chemical Engineering Science, 61*(3), 1027-1040.

Yan, Y., Zhu, X., Yu, Y., Li, C., Zhang, Z., and Wang, F. (2022). Nanotechnology strategies for plant genetic engineering. *Advanced Materials, 34*(7), 2106945.

Yang, A., Su, Q., and An, L. (2009). Ovary-drip transformation: a simple method for directly generating vector-and marker-free transgenic maize (*Zea mays* L.) with a linear GFP cassette transformation. *Planta*, 229, 793-801.

Zaidi, S. S. E. A., and Mansoor, S. (2017). Viral vectors for plant genome engineering. *Frontiers in Plant Science, 8*, 539.

Zhang, L., Gu, L., Ringler, P., Smith, S., Rushton, P. J., and Shen, Q. J. (2015). Three WRKY transcription factors additively repress abscisic acid and gibberellin signaling in aleurone cells. *Plant Science*, 236, 214-222.

Zhang, Y., Malzahn, A. A., Sretenovic, S., and Qi, Y. (2019). The emerging and uncultivated potential of CRISPR technology in plant science. *Nature Plants*, 5(8), 778-794.

Zhao, X., Meng, Z., Wang, Y., Chen, W., Sun, C., Cui, B.,. and Cui, H. (2017). Pollen magnetofection for genetic modification with magnetic nanoparticles as gene carriers. *Nature plants*, 3(12), 956-964.

Zhou, M., Luo, J., Xiao, D., Wang, A., He, L., and Zhan, J. (2023). An efficient method for the production of transgenic peanut plants by pollen tube transformation mediated by *Agrobacterium tumefaciens*. *Plant Cell, Tissue and Organ Culture (PCTOC)*, 152(1), 207-214.

Chapter 3

Trends in Molecular Markers: Current Status and Application

Parinita Borah and Pooja Kumari*

Ph.D. Scholar, Department of Plant Breeding and Genetics, Assam Agricultural University, Jorhat – 785 013, Assam
**e-mail: porinita05@gmail.com*

ABSTRACT

Molecular markers, which encompass genetic, epigenetic, and proteomic variations, have emerged as indispensable tools in modern biological and medical research. The chapter begins by elucidating the fundamental concepts of molecular markers, emphasizing their significance in characterising genetic diversity, understanding evolutionary relationships, and unravelling complex biological processes. It delves into the intricate mechanisms of various molecular marker technologies, ranging from classic DNA-based markers like microsatellites and amplified fragment length polymorphisms (AFLPs) to the more recent developments in next-generation sequencing (NGS) and high-throughput genotyping arrays. The advent of NGS technologies has enabled comprehensive genome-wide analyses, opening avenues for identifying genetic variants associated with diseases, drug responses, and traits of interest of crop. Molecular markers have catalysed advancements in crop improvement and breeding strategies. Marker-assisted selection (MAS) expedites the identification of plants or animals carrying desirable traits, accelerating the breeding process and enabling the development of high-yield, disease-resistant, and climate-resilient varieties.

Keywords: Marker, MAS, CRISPR/Cas9, Genomic selection, GWAS.

INTRODUCTION

Molecular markers have transformed the landscape of genetics and biological research by providing valuable insights into genetic diversity, evolutionary relationships, and various applications. Molecular markers are important tools in various fields of biological and medical research, including genetics, genomics, medicine, and agriculture. They are used to identify specific DNA or protein sequences that can help researchers to understand thev genetic variation, track the inheritance of traits, diagnose diseases, and even guide personalized treatment strategies. (Jiang.,2021)

We can better understand the molecular underpinnings of numerous biological events in plants by finding and analysing genetic variation. Since sequencing initiatives cannot be used to study the whole plant world, molecular markers and their relationships to phenotypes give us the necessary landmarks to understand genetic diversity. Genetic or DNA based marker techniques such as RFLP (restriction fragment length polymorphism), RAPD (random amplified polymorphic DNA), SSR (simple sequence repeats) and AFLP (amplified fragment length polymorphism) are routinely being used in ecological, evolutionary, taxonomical, phylogenic and genetic studies of plant sciences. (Agarwal *et al.*, 2008). Recently, the advent of the next-generation sequencing (NGS) technologies has contributed to highly abundant and informative markers like single-nucleotide polymorphisms (SNPs). The development of saturated linkage maps could only be possible with the availability of molecular markers. These maps are prerequisite for gene/QTL mapping, map-based cloning of genes, and marker-assisted selection. This chapter briefly discusses the different types of plant molecular markers while particularly focusing on recent developments in molecular marker technologies. (Bachlava *et al.*, 2022)

Definition of Molecular Marker

Molecular markers are specific DNA sequences or genetic variations that can be used to identify unique individuals, populations, or species. These markers serve as genetic landmarks, allowing researchers to distinguish between different organisms based on their genetic profile.

Scientists can learn more about the genetic variety found in various groups of animals by examining the presence or absence of certain markers. Understanding how populations adjust to shifting surroundings and how species change through time requires knowledge of this information. Molecular markers play a pivotal role in population genetics studies. They help researchers decipher the genetic structure of populations, investigate gene flow and migration patterns, and identify potential barriers to gene exchange. This knowledge contributes to our understanding of species distribution, conservation strategies, and evolutionary dynamics. In agriculture, molecular markers are used for marker-assisted breeding. By identifying markers associated with desirable traits, such as disease resistance or high yield, breeders can select plants or animals with these markers, leading to more efficient and targeted breeding programs. It also used in forensic science for human identification and criminal investigations. Techniques like DNA fingerprinting and DNA profiling rely on the uniqueness of certain genetic markers within an individual's genome. Molecular markers provide data for constructing phylogenetic trees and evolutionary studies among species. By comparing the presence and absence of markers across taxa, researchers can infer evolutionary histories and timelines.

Historical Overview of Molecular Markers

Since the early 20th century, when researchers first started examining how certain genes are associated with apparent features, the idea of using genetic markers to analyse variation and relatedness has been around. However, it wasn't until the development of molecular biology techniques that researchers could explore genetic markers at the DNA level. One of the earliest breakthroughs was the discovery of restriction enzymes in the 1970s. These enzymes could cut DNA at specific recognition sites, producing unique DNA fragments. This discovery led to the creation of the RFLP technique, which involved digesting DNA with these enzymes, separating the resulting fragments, and probing them with radioactive DNA fragments to visualize specific genetic regions.

As technology advanced, other marker techniques emerged. The Random Amplified Polymorphic DNA (RAPD) method, developed in the 1980s, allowed for the amplification of random DNA fragments using short, arbitrary primers. This technique provided a quicker and simpler way to analyse genetic variation. The advent of DNA sequencing in the 1990s opened new avenues for molecular marker development. Microsatellites, short repeated DNA sequences, gained popularity due to their high variability and usefulness in population genetics studies. Next-generation sequencing (NGS) technologies revolutionized(2000s) the acquisition of genetic data, enabling rapid and cost-effective sequencing of entire genomes. NGS paved the way for genome-wide association studies (GWAS), where molecular markers were used to correlate genetic variations with diseases and traits. The 2010s were dominated by the prominence of single nucleotide polymorphisms (SNPs) as the cornerstone of molecular marker research. High-throughput genotyping platforms facilitated the analysis of vast numbers of SNPs, leading to advances in personalized medicine, pharmacogenomics, and ancestry tracing. These markers facilitated the translation of genetic insights into clinically relevant applications.

Types of Molecular Markers

Molecular markers are tools used to detect genetic variations at the DNA level. They aid in various genetic, genomic, and breeding studies. Here are some of the main types of molecular markers

Classical Markers

Morphological Marker

Morphological markers are a type of observable trait used in genetics and biology to characterize and differentiate organisms based on their physical appearance. These markers are particularly valuable in fields such as taxonomy, evolution, ecology, and plant and animal breeding, where variations in outward characteristics are used to identify species, assess genetic diversity, and make informed decisions about breeding programs. Unlike molecular markers that involve analysing genetic material at the molecular level, morphological markers are based on visible traits that can be easily observed and measured. Morphological markers encompass a wide range of physical attributes, including size, shape, color, pattern, structure, and other visible traits that are externally discernible. These traits can pertain to different parts of an organism, such as leaves, flowers, fruits, seeds, and overall body structure.

Cytological Markers

Cytological markers, also known as cytogenetic markers, are used in plant breeding to detect and characterize structural or numerical changes in chromosomes. These changes can serve as indicators of genetic diversity, chromosomal aberrations, and specific traits in plants. Cytological markers play a crucial role in understanding the genetic makeup of plants, assisting in breeding programs, and facilitating the selection of desirable traits. These markers provide insights into the genetic and evolutionary relationships among different plant species and varieties. Some stain like sGiemsa stain produces G bands, quinacrine hydrochloride produces Q bands, and reversed G bands are known as R bands (Jiang.,2012). These chromosomal landmarks are commonly employed in physical mapping and linkage group identification in addition to being used to characterise normal chromosomes and detect chromosome mutations. Cytological markers involve the study of chromosome structure, number, and behaviour. This includes identifying variations such as chromosomal translocations, inversions, deletions, duplications, and aneuploidy (abnormal chromosome numbers). These variations can impact plant traits and performance.

Biochemical/Protein Markers

Biochemical markers are compounds or molecules present in plants that can be quantified and linked to specific genetic traits. These markers are often associated with metabolic processes, and their presence or concentration can reflect the expression of certain genes. They are particularly useful in identifying genetic variations that might not be visually apparent. Some common types of biochemical markers include. Isozymes are different forms of enzymes that catalyze the same reaction but are encoded by different genes. They can be separated and distinguished based on their electrophoretic mobility, revealing genetic variation that might not be apparent from external traits. These secondary metabolites are involved in plant defense mechanisms and have diverse roles in growth and development. Their concentrations can vary significantly between different plant varieties and can serve as markers for specific traits.

Molecular Markers/DNA Markers

A molecular marker is a specific and easily detectable DNA sequence that is associated with a particular trait, characteristic, or genetic condition in an organism. Molecular markers are used in various fields of biology, including genetics, genomics, and plant and animal breeding, to identify and track the presence or absence of specific genetic trait. By increasing genome coverage, the improved marker approaches additionally make use of emerging classes of DNA elements such retrotransposons, mitochondrial, and chloroplast based microsatellites to show genetic diversity (Agarwal *et al.*, 2008).

Classification of molecular markers are classified into various groups on the basis of:

1. Mode of gene action (co-dominant or dominant markers);
2. Method of detection (hybridization-based molecular markers or polymerase chain reaction (PCR)-based markers);
3. Mode of transmission (paternal organelle inheritance, maternal organelle inheritance, bi-parental nuclear inheritance or maternal nuclear inheritance) (Baloch *et al.*, 2017).

Hybridization-Based Markers (RFLP)

The Restriction Fragment Length Polymorphism (RFLP) marker is one of the pioneering molecular tools used in genetics and genomics. RFLPs are variations in DNA sequences recognized by the action of specific enzymes known as restriction endonucleases. These enzymes cleave DNA at specific recognition sites, resulting in the formation of DNA fragments of varying lengths.

PCR Based Marker

There are several PCR-based methods that may be used to find polymorphisms in plants. It is crucial that these marker technologies can be shared between laboratories for their widespread use in germplasm characterisation and breeding. To do this, they must also be standardised to produce repeatable results, allowing for the direct collation and comparison of the data (Jones *et al.*, 1997)

Randomly Amplified Polymorphic DNA (RAPD)

The Random Amplified Polymorphic DNA (RAPD) marker is a molecular biology technique used to detect genetic variations within individuals or populations. RAPD markers are based on the polymerase chain reaction (PCR) amplification of random segments of DNA using short, single arbitrary primers.

This method does not require prior knowledge of the DNA sequence, making it a simple and cost-effective way to identify DNA polymorphisms.

RAPD markers are well suited for genetic mapping, for plant and animal breeding applications, and for DNA fingerprinting, with particular utility for studies of population genetics (William *et al.*, 2018).

AFLP

Amplified Fragment Length Polymorphism (AFLP) is a powerful molecular marker technique used for analyzing genetic diversity, population genetics, and evolutionary relationships among organisms. Gene expression fingerprints have also been created using AFLP-based techniques (Chial.,2008). AFLP combines aspects of restriction enzyme digestion and PCR amplification to generate a fingerprint-like pattern of DNA fragments, which reflects variations in the genome. AFLP markers are employed in genetic mapping, population genetics, phylogenetic analysis, and studies of genetic diversity. They are especially valuable for non-model organisms. AFLP markers are generally considered dominant, meaning they don't provide information about heterozygosity. The innovative molecular fingerprinting approach known as amplified fragment length polymorphism (AFLP) may be used to identify DNA from any source and of any complexity (Blears *et al.*, 1998)

SSRs or Microsatellites

Simple Sequence Repeats, commonly known as SSRs or microsatellites, are small repeating sequences of DNA that are scattered throughout an organism's genome. These repetitive motifs are short, typically consisting of 1 to 6 nucleotide units, and are repeated in tandem. SSRs are found in both coding and non-coding regions of the genome and have been extensively used as genetic markers due to their high degree of polymorphism and co-dominant inheritance pattern. SSRs are highly polymorphic, meaning they exhibit variations in the number of repeats among individuals. This makes them invaluable for studying genetic diversity, population genetics, and inheritance patterns.

Chloroplast Microsatellites (cpSSRs)

Chloroplast microsatellites, also known as cpSSRs (chloroplast simple sequence repeats) or chloroplast SSRs, are a type of genetic marker found within the chloroplast DNA of plants. Just like nuclear microsatellites, chloroplast microsatellites consist of short, repetitive DNA sequences, but they are located in the chloroplast genome rather than the nuclear genome. These markers have proven to be valuable tools for studying plant genetics, population dynamics, and evolutionary relationships.sw Chloroplast microsatellites are highly variable among plant species and even among individuals within a species. They provide insights into the genetic diversity and structure of plant populations, which is essential for conservation efforts and understanding how plant species adapt to their environments.

Mitochondrial Microsatellites

Mitochondrial DNA is inherited maternally, meaning it is passed from mother to offspring. This unique mode of inheritance makes mitochondrial markers particularly useful for studying maternal lineages, as well as migration and historical movement of populations. Additionally, the high mutation rate of mitochondrial DNA makes it a valuable tool for tracing evolutionary histories and understanding ancient relationships. Mitochondrial microsatellites are employed to study maternal ancestry and lineage. These markers can help trace the origins of populations and track their migratory routes over time.

Randomly Amplified Microsatellite Polymorphisms (RAMP)

RAMP, also known as Randomly Amplified Microsatellite Polymorphisms, is a technique used to identify genetic variations within a population. It combines the principles of random amplification, as seen in RAPD (Random Amplified Polymorphic DNA), with the focus on microsatellite regions. Microsatellites are short tandem repeats of DNA sequences that are highly variable between individuals due to variations in the number of repeat units. RAMP can provide insights into the evolutionary relationships and divergence among populations or species by analysing the shared and unique polymorphisms.

Sequence-Related Amplified Polymorphism (SRAP)

Sequence-Related Amplified Polymorphism (SRAP) is a molecular marker technique used to detect genetic variation within a population. SRAP combines aspects of polymerase chain reaction (PCR) amplification with the analysis of specific DNA regions, allowing researchers to uncover polymorphisms in genomic DNA. SRAP is particularly useful in studies of genetic diversity, population genetics, and plant breeding.

Inter Simple Sequence Repeat (ISSR)

Inter Simple Sequence Repeat (ISSR) is a molecular marker technique commonly used in genetics and genomics research to analyse genetic diversity, relationships between individuals, and population dynamics. ISSR markers are based on the amplification of regions between two microsatellite sequences, providing insights into the genetic makeup of various organisms. ISSR markers rely on the presence of microsatellite regions, which are short tandem repeats of DNA sequences. Unlike other techniques that amplify between a primer and a microsatellite sequence, ISSR uses two primers that bind to microsatellite regions located on opposite DNA strands, amplifying the region in between. The amplified fragments are usually polymorphic due to variations in the number of repeats within microsatellite sequences.

Retrotransposons

Retrotransposons are a class of mobile genetic elements that play a significant role in shaping the genomes of various organisms. They are often used as genetic markers to study evolutionary relationships, population dynamics, and genetic diversity. Retrotransposons, also known as transposable elements, have contributed to genome evolution by their ability to copy and insert themselves into different locations within a genome.

There are two main classes of retrotransposons:

1. **Long Terminal Repeat (LTR) Retrotransposons:** These retrotransposons have long terminal repeat sequences at their ends and can resemble retroviruses in their structure. They contain genes for reverse transcriptase and integrase enzymes necessary for their replication and insertion.

2. **Non-LTR Retrotransposons:** This group includes two subtypes: LINEs (Long Interspersed Elements) and SINEs (Short Interspersed Elements). Non-LTR retrotransposons lack terminal repeats but contain their own reverse transcriptase. They are less complex than LTR retrotransposons but still capable of autonomous replication and insertion.

Retrotransposon-Based Insertion Polymorphism (RBIP)

Retrotransposon-Based Insertion Polymorphism (RBIP) is a genetic marker technique that utilizes the presence or absence of retrotransposon insertions in genomes to study genetic diversity, population dynamics, and evolutionary relationships. RBIP leverages the natural variability in the presence of retrotransposons at specific genomic locations to uncover insights into the genetic makeup of different individuals and populations. Retrotransposons are mobile genetic elements that can move around a genome via a copy-and-paste mechanism. RBIP focuses on specific retrotransposon insertions in different individuals or populations. The presence or absence of these insertions creates a type of genetic polymorphism that can be used as a marker for genetic studies

Inter-Primer Binding Site (iPBS)

The Inter-Primer Binding Site (iPBS) is a concept and technique used to study and analyze the presence or absence of specific retrotransposon insertions in genomic DNA. iPBS focuses on the regions flanking a retrotransposon insertion site, specifically the binding sites of primers designed to amplify the DNA segment containing the retrotransposon. This approach allows researchers to identify and analyze the variation in retrotransposon insertions among different individuals or populations. Retrotransposons are mobile genetic elements that can insert themselves into various genomic locations. iPBS leverages this natural variability by designing primers that target the genomic regions immediately adjacent to the retrotransposon insertion site. The presence of the retrotransposon insertion disrupts the primer binding site, resulting in variations in the PCR amplification of the region.

Cleaved Amplified Polymorphic Sequences (CAPS)

Cleaved Amplified Polymorphic Sequences (CAPS) is a molecular marker technique widely used in genetics and genomics research to detect single nucleotide polymorphisms (SNPs) and small insertions or deletions (indels) within DNA sequences. CAPS leverages the presence of restriction enzyme recognition sites that are affected by sequence variations, allowing for the differentiation of alleles based on the digestion pattern of amplified DNA fragments. The CAPS technique relies on the fact that specific restriction enzymes recognize and cleave DNA sequences at specific recognition sites. If a genetic variation, such as a SNP or indel, disrupts or creates a recognition site, it can lead to differences in the DNA digestion pattern. By designing primers that flank the target region containing the SNP or indel, amplifying the DNA, and subsequently digesting it with a restriction enzyme, CAPS enables the detection of sequence variations.

Sequence-Characterized Amplified Regions (SCAR)

Sequence-Characterized Amplified Regions (SCAR) is a molecular marker technique used in genetics and genomics research to target and amplify specific DNA sequences of interest. SCAR markers are derived from DNA fragments that have been amplified using Polymerase Chain Reaction (PCR) with specific primers designed based on known sequences. These markers find application in various genetic studies, including genetic diversity assessment, trait mapping, and evolutionary analyses. The central principle of SCAR involves the selective amplification of DNA regions that are associated with specific traits, genetic markers, or sequences of interest. Unlike some other molecular marker techniques that rely on random genomic sequences, SCAR markers are designed based on known sequences or regions associated with particular traits.

Sequence-Based Markers

Sequence-based markers are a class of molecular markers used in genetics and genomics research to identify and analyze specific DNA sequences within an organism's genome. These markers rely on the variation in DNA sequences, such as single nucleotide polymorphisms (SNPs) or insertion-deletion (indel) events, to study genetic diversity, evolutionary relationships, and traits. Unlike some other marker types that rely on techniques like PCR and gel electrophoresis, sequence-based markers involve the direct sequencing of DNA to reveal variations. a significant proportion of polymorphic SNPs and a high degree of interspecies transferability. In fact, polymorphism was found in more than 95 per cent of effective SNP tests, and more than 90 per cent of these assays could be successfully transferred to closely related wild species

Sanger Method of Sequencing

The Sanger method of DNA sequencing, also known as chain termination or dideoxy sequencing, is a pioneering technique that revolutionized genetics and genomics research. Developed by Frederick Sanger and colleagues in the late 1970s, this method played a pivotal role in deciphering the human genome and laid the foundation for modern DNA sequencing technologies. The Sanger method is based on DNA replication using DNA polymerase, with the incorporation of modified nucleotides that lack a 3'-OH group (dideoxynucleotides or ddNTPs). When a ddNTP is incorporated into the growing DNA strand, it terminates the chain elongation process. By performing separate reactions for each of the four ddNTPs (A, C, G, T), labeled with distinct fluorescent dyes, the DNA fragments are terminated at every position corresponding to the specific base

Pyrosequencing

Pyrosequencing-based markers are a type of DNA sequencing technique that enables the rapid determination of DNA sequences by measuring the release of light generated during DNA synthesis. This method provides a cost-effective and high-throughput approach to analyzing genetic variations, making it valuable for genetic research, clinical diagnostics, and personalized medicine.Pyrosequencing is based on the detection of light emitted when a nucleotide is incorporated into a growing DNA strand. This method relies on the detection of pyrophosphate (PPi) released during DNA synthesis. As DNA polymerase incorporates a nucleotide, it releases PPi, which is then converted to ATP by the enzyme ATP sulfurylase. Luciferase enzyme further catalyzes the conversion of ATP to light. The amount of light generated is proportional to the number of nucleotides incorporated, allowing the sequence of the DNA strand to be determined

Next-Generation Sequencing (NGS)

Next-Generation Sequencing (NGS), also known as high-throughput sequencing, is a revolutionary DNA sequencing technology that has transformed genomics research and applications. NGS enables the rapid and cost-effective analysis of large volumes of DNA or RNA sequences, allowing researchers to uncover genetic variations, study gene expression, and explore the complexities of genomes on a scale previously unattainable. NGS methods employ massively parallel sequencing, which means that multiple DNA fragments are sequenced simultaneously. Unlike traditional Sanger sequencing, where a single DNA strand is sequenced sequentially, NGS platforms can generate millions to billions of short reads in a single run. These reads are then computationally assembled to reconstruct the entire genomic or transcriptomic sequence.

Genotyping by Sequencing (GBS)

Genotyping by sequencing (GBS) is a high-throughput DNA sequencing technique used to identify genetic variations, particularly single nucleotide polymorphisms (SNPs), within genomes. GBS combines DNA sample preparation, next-generation sequencing (NGS), and bioinformatics analysis to rapidly analyse large numbers of individuals, making it valuable for studying genetic diversity, population genetics, and trait associations. New genetic markers for complex and unstructured populations have been made possible by next-generation sequencing techniques such as genotyping-by-sequencing and association mapping(Nadeem *et al.*, 20017). The principle of GBS involves selectively sequencing specific regions of a genome to identify genetic variations. Instead of sequencing the entire genome, GBS focuses on targeted regions, reducing the sequencing complexity and cost while still capturing a representative portion of genetic diversity.

Single-Nucleotide Polymorphism (SNP)

A Single-Nucleotide Polymorphism (SNP) is one of the most common types of genetic variation found in DNA sequences. SNPs are single-base differences in the DNA sequence that occur at specific positions in the genome. These variations contribute to the genetic diversity among individuals and play a crucial role in disease susceptibility, traits, evolution, and personalized medicine. SNPs result from single-nucleotide changes in the DNA sequence. They occur when one of the four nucleotides (adenine, cytosine, guanine, or thymine) is replaced by another at a specific position in the genome. For example, if a DNA sequence originally has an adenine (A) at a certain position, a SNP may result in that adenine being replaced by a cytosine (C), guanine (G), or thymine (T). a significant proportion of polymorphic SNPs and a high degree of interspecies transferability. In fact, polymorphism was found in more than 95 per cent of effective SNP tests, and more than 90 per cent of these assays could be successfully transferred to closely related wild species (Bachalva *et al.*, 2012).

Diversity Array Technology (DArT Seq)

Diversity Array Technology (DArT-Seq) is a high-throughput genotyping technique used to analyse genetic diversity and polymorphisms within genomes. DArT-Seq combines elements of microarray technology and next-generation sequencing (NGS) to efficiently and comprehensively study thousands of genetic markers across multiple individuals or populations. This technology is particularly useful for genetic diversity studies, marker-assisted breeding, and population genomics. The principle of DArT-Seq involves the simultaneous analysis of thousands of genetic markers distributed throughout the genome. It encompasses two main steps: complexity reduction and sequencing.

RNA-Based Markers

RNA-based markers are genetic markers that utilize information from RNA molecules to analyse gene expression, alternative splicing, and other RNA-related processes. These markers provide insights into the dynamic regulation of genes and their functional activities within cells and tissues. RNA-based markers play a crucial role in various biological and medical research applications, including understanding disease mechanisms, characterizing cellular responses, and identifying potential therapeutic targets.

Inter Small RNA Polymorphism

Inter-species small RNA polymorphism refers to the genetic variations or differences observed in small RNA molecules across different species. Small RNAs are short RNA molecules, typically ranging from about 20 to 30 nucleotides in length, that play significant roles in gene regulation by targeting specific mRNA molecules for degradation or translational repression.Small RNA molecules include various types, such as microRNAs (miRNAs), small interfering RNAs (siRNAs), and piwi-interacting RNAs (piRNAs), each with specific functions and mechanisms of action. These small RNAs are known to be conserved across species to varying degrees, meaning that similar small RNA sequences and functions are found in diverse organisms.

EST-SSR

An Expressed Sequence Tag (EST) is a short DNA sequence derived from a fragment of a gene that is actively being transcribed and translated into a protein. The possibility to assess the number and relative distribution of microsatellites to across transcribed and nontranscribed areas, as well as the link between these properties and haploid genome size, is made possible by large-scale genomic and EST sequencing(Morgante *et al.*, 2002). ESTs are obtained through a process called cDNA (complementary DNA) sequencing, which involves converting the mRNA (messenger RNA) produced by a specific gene into complementary DNA and then sequencing this cDNA. ESTs provide valuable insights into the genes that are actively expressed in a specific tissue or under certain conditions, and they help researchers understand the genetic basis of various biological processes.

Current Status of Molecular Marker Technologies

The landscape of molecular marker technologies has evolved rapidly in recent years, with significant advancements in DNA sequencing, high-throughput genotyping platforms, Next-Generation Sequencing (NGS) applications, and Genome-Wide Association Studies (GWAS). These technologies have revolutionized our ability to analyse genetic variation, study complex traits, and understand the genetic underpinnings of various phenomena. In this section, we'll explore the current status and impact of these molecular marker technologies.

Advancements in DNA Sequencing

The advent of high-throughput DNA sequencing has been a game-changer in molecular marker research. Traditional Sanger sequencing, while valuable, was limited in its throughput and cost-effectiveness. Next-generation sequencing (NGS) technologies, including Illumina, Ion Torrent, and Pacific

Biosciences, have dramatically increased sequencing speed and reduced costs. This has enabled researchers to sequence entire genomes or targeted regions at unprecedented scales.

Key Developments

1. **Whole Genome Sequencing (WGS):** NGS allows the sequencing of entire genomes, providing a comprehensive view of an organism's genetic makeup. WGS has become more accessible, enabling studies in diverse species and populations.

2. **Targeted Sequencing:** Researchers can selectively sequence specific genomic regions using techniques like exome sequencing (targeting protein-coding regions) or amplicon sequencing (targeting specific DNA regions). This approach is cost-effective and well-suited for studies focusing on functional elements.

3. **Metagenomic Sequencing:** NGS has empowered the exploration of microbial communities in diverse environments through metagenomic sequencing. This has implications in environmental studies, microbiome research, and more.

Impact

☆ **Diverse Applications:** DNA sequencing has expanded beyond traditional marker analysis, allowing comprehensive genome-wide studies, mutation detection, and functional genomics.

☆ **Population Genomics:** Large-scale sequencing facilitates population-level analyses, revealing genetic diversity, adaptation, and migration patterns.

High-Throughput Genotyping Platforms

High-throughput genotyping platforms have made genotyping large numbers of markers across multiple samples faster and more cost-effective. These platforms utilize microarray technology or multiplex PCR to simultaneously interrogate thousands to millions of genetic markers.

Key Platforms

1. **SNP Arrays:** SNP arrays, such as those from Affymetrix and Illumina, are widely used for genotyping single nucleotide polymorphisms (SNPs) on a large scale. These arrays enable genome-wide association studies (GWAS) and linkage analysis.

2. **Genotyping by Sequencing (GBS):** GBS combines next-generation sequencing with genotyping, allowing cost-effective genotyping of thousands of markers in a targeted manner.

3. **Multiplex PCR:** Multiplex PCR assays, often using fluorophore labelled primers, enable targeted genotyping of specific markers, making it a valuable tool in diverse applications.

Impact

- ☆ **Speed and Efficiency:** High-throughput genotyping platforms significantly accelerate the genotyping process, making it feasible to analyse thousands of markers in large sample sizes.

- ☆ **Disease Studies:** These platforms have enabled GWAS, helps to the discovery of genetic variantion associated with different complex diseases and traits.

- ☆ **Breeding Programs:** High-throughput genotyping facilitates marker-assisted breeding by allowing breeders to select individuals with desired traits more efficiently.

Next-Generation Sequencing (NGS) Applications

Next-generation sequencing (NGS) has revolutionized molecular marker research by enabling massive parallel sequencing of DNA or RNA molecules. The scalability, speed, and cost-effectiveness of NGS have opened up new avenues for diverse applications.

Key NGS Applications

1. **Transcriptomics (RNA-Seq):** RNA-Seq allows researchers to quantify gene expression levels across the transcriptome. This is essential for understanding gene regulation, identifying differentially expressed genes, and studying gene function.

2. **Epigenomics:** NGS can be used to study epigenetic modifications, such as DNA methylation and histone modifications, gene regulation and chromatin structure.

3. **ChIP-Seq:** Chromatin Immunoprecipitation sequencing (ChIP-Seq) enables the genome-wide mapping of DNA-protein interactions, shedding light on transcription factor binding sites and histone modifications.

4. **Metagenomics:** NGS has transformed metagenomic studies by enabling the comprehensive analysis of microbial communities in various environments, including the human microbiome and environmental samples.

Impact

- ☆ **Holistic Insights:** NGS applications provide a comprehensive understanding of gene expression, epigenetic modifications, and DNA-protein interactions, facilitating functional genomics studies.

- ☆ **Discovery of Regulatory Elements:** Techniques like ChIP-Seq reveal essential regulatory elements in the genome, contributing to our understanding of gene regulation.
- ☆ **Microbiome Studies:** NGS has been pivotal in studying microbial communities, aiding in understanding their role in health, ecology, and disease.

Genome-Wide Association Studies (GWAS)

GWAS have become a cornerstone of molecular genetics, aiming to identify genetic variants associated with complex traits, diseases, or phenotypes. These studies leverage molecular markers to scan the entire genome for associations between genetic variations and specific traits.

Key Aspects of GWAS

1. **Marker Density:** GWAS typically use high-density SNP arrays to genotype thousands to millions of SNPs across the genome. This comprehensive genotyping approach allows the identification of associations even in regions not previously linked to the trait of interest.

2. **Statistical Analysis:** Advanced statistical techniques are used to find connections between characteristics and markers that are meaningful, taking population structure, genetic relatedness, and multiple testing corrections into account.

3. **Replication and Validation:** GWAS findings require replication in independent populations to establish the robustness of the identified associations. Validation studies using different cohorts help confirm the significance of the identified markers.

Impact

- ☆ **Disease Genetics:** GWAS have revolutionized our understanding of the genetic basis of complex diseases, identifying thousands of genetic loci associated with traits such as diabetes, cardiovascular diseases, and autoimmune disorders.
- ☆ **Pharmacogenomics:** GWAS have implications for personalized medicine by identifying genetic markers that influence drug responses.
- ☆ **Trait Discovery:** Beyond diseases, GWAS have been instrumental in discovering genetic variants associated with various traits, including height, cognitive abilities, and longevity.

The current status of molecular marker technologies reflects a dynamic and rapidly evolving field. Advancements in DNA sequencing, high-throughput genotyping platforms, NGS applications, and GWAS have expanded our capabilities to analyse genetic variation, study complex traits, and unravel the genetic underpinnings of various phenomena. These technologies continue to drive groundbreaking discoveries in genetics, genomics, and related fields, with profound implications for personalized medicine, conservation, agriculture, and our understanding of the natural world.

Applications of Molecular Markers

Molecular markers are essential tools in various fields of biology, genetics, and agriculture. They are specific DNA sequences that can be easily detected and used to identify individuals, track genetic variation, and analyze relationships between different organisms. Molecular markers have numerous applications across different disciplines, including medicine, forensics, evolutionary biology, and agriculture

Cultivar Identity

Molecular markers are segments of DNA that are used to identify specific genetic traits or variations within individuals, populations, or species. These markers serve as unique signposts within the genome, allowing scientists to track genetic diversity, inheritance patterns, evolutionary relationships, and various other genetic characteristics. Molecular markers have become indispensable tools in genetics, genomics, and related fields due to their ability to provide insights into the genetic makeup of organisms. There are several types of molecular markers commonly used in genetic research such as SSR, SNP, RFLP, AFLP, RAPD *etc.*

Study of Heterosis

Molecular markers play a crucial role in studying heterosis, also known as hybrid vigor, which is the phenomenon where the offspring of two genetically diverse parents exhibit superior traits compared to their parents. Understanding the genetic basis of heterosis can have significant implications for plant and animal breeding, as it can lead to the development of improved hybrid varieties with enhanced traits such as yield, disease resistance, and stress tolerance. Molecular markers contribute to study of heterosis through Marker-Assisted Selection (MAS) for Parental Line Development, Genetic Mapping and QTL Analysis, Marker-Based Prediction of Hybrid Performance,

Identification of Genomic Region Selection

Molecular markers play a crucial role in identifying genomic regions under selection, a process known as selective sweep detection. Selective sweeps occur when a beneficial allele or a set of alleles rapidly increases in frequency within a population due to natural selection. Detecting these regions provides insights into evolutionary processes, adaptation, and the genetic basis of traits. Molecular markers allow researchers to compare allele frequencies between different populations or groups. Genomic regions that show significantly different allele frequencies are potential candidates for selection. Methods such as F-statistics (FST) and principal component analysis (PCA) can help quantify genetic differentiation and identify outlier loci that deviate from neutral expectations.

Assessment of Genetic Diversity and Parental Selection

Molecular markers are invaluable tools for assessing genetic diversity and facilitating parental selection in plant breeding programs. Genetic diversity analysis is crucial for selecting appropriate parents that will contribute to the development of improved plant varieties. Here's how molecular markers help in these processes. Molecular markers create unique genetic profiles for individual plants. By comparing these profiles, breeders can assess the genetic diversity within a population. This information helps in selecting diverse parents for hybridization. Molecular markers allow the identification of subpopulations within a larger population.

Marker-Assisted Introgression

Marker-assisted introgression is a breeding technique that uses molecular markers to facilitate the transfer of specific desirable genes or genomic regions from one parental line (donor) into another parental line (recipient) to create improved plant varieties. This technique is particularly useful when breeding for complex traits such as disease resistance, stress tolerance, or nutritional content. Marker-assisted introgression combines traditional breeding methods with molecular tools to streamline the selection process and increase the efficiency of transferring targeted genetic material. An abundance of DNA marker-trait connections has been produced by the numerous quantitative trait loci (QTLs) mapping studies conducted for several crop species. (Collard and Mackill, 2007).

Selecting the Donor and Recipient Parents: The first step is to choose a donor parent that possesses the desired trait or gene of interest. This parent typically originates from a wild relative or a different variety with the desired characteristic. The recipient parent is the cultivated variety that will receive the beneficial genetic material.

Identifying Molecular Markers Linked to the Trait: Molecular markers associated with the trait or gene of interest are identified through genetic mapping or other molecular techniques. These markers are located near the genomic region containing the desired trait.

Marker-Assisted Selection (MAS): The recipient parent is crossed with the donor parent to produce hybrid offspring. These hybrids are then screened for the presence of the molecular markers linked to the desired trait. This step allows breeders to select individuals that have inherited the target genomic region from the donor parent.

Backcrossing: The selected hybrid individuals, which carry the desired genomic region, are then backcrossed with the recipient parent multiple times. Backcrossing helps dilute the genetic background of the donor parent and recover the genetic background of the recipient parent while retaining the trait of interest.

Marker-Based Selection during Backcrossing: Molecular markers are used at each backcross generation to track the presence of the target genomic region. This helps breeders ensure that the trait is being introgressed while minimizing the retention of unwanted donor genetic material.

Phenotypic Evaluation: Throughout the backcrossing process, individuals are evaluated for the expression of the desired trait. This helps confirm that the trait of interest has been successfully introgressed from the donor parent. Some example has shown in the Table 3.1.

Table 3.1: The Marker Assisted Backcrossing in different Crops

Species	Trait	Gene/QTLs	Foreground selection	Background selection
Barley	Barley yellow dwarf virus	Yd2	STS	Not performed
	Leaf rust	Rphq6	AFLP	AFLP
	Stripe rust	QTLs on 4H and 5H	Not performed	Not performed
	Yield	QTLs on 2H and 3HL	RFLP	RFLP
Maize	Corn borer resistance	QTLs on chromosome 7, 9 and 10	RFLP	RFLP
	Earliness and yield	QTLs on chromosome 5, 8 and 10	RFLP	RFLP
Rice	Early blight	Xa21	STS[a]	RFLP
	Early blight	Xa21	STS[a]	AFLP
	Early blight	Xa5, xa13 and xa21	STS, CAPS	Not performed
	Blast	Pi1	SSR	ISSR[b]
	Deep roots	QTLs on chromosome 1, 2, 7, and 9	RFLP and SSR	SSR
	Tolerance, disease, resistance, quality	Subchr9 QTL, Xa21, Bph and blast QTLs, and quality loci	SSR and STS	Not performed
Wheat	Powdery mildew	22 Pm genes	Phenotyping	AFLP

Evolution and Phylogeny

Molecular markers play a significant role in studying evolution and phylogeny (the evolutionary relationships between different species or groups). They provide insights into the genetic changes, relationships, and divergence times among organisms. Here's how molecular markers help in the study of evolution and phylogeny. Molecular markers involve analysing the DNA sequences of specific genes or regions in different species. By comparing these sequences, researchers can identify similarities and differences, which help determine the evolutionary relationships between species. These are fundamental for constructing phylogenetic trees, which depict the branching patterns of evolutionary relationships. DNA sequence data from markers are used to infer the sequence of divergence events, revealing the most likely evolutionary history.

Marker-Assisted Pyramiding

Marker-assisted pyramiding is a breeding strategy used to combine multiple beneficial genes or traits into a single plant variety. This approach aims to enhance the performance and resilience of the resulting variety by stacking multiple desired traits from different parental lines. The term "pyramiding" refers to the accumulation or stacking of these traits in a single genetic background. Molecular markers are employed to facilitate the selection of plants carrying the target traits during the breeding process. Marker-assisted pyramiding is widely used in modern plant breeding to develop superior crop varieties that can address multiple challenges simultaneously, such as pests, diseases, environmental stressors, and changing consumer preferences. It exemplifies how molecular tools can significantly accelerate and improve the efficiency of trait incorporation into commercial crop varieties shown in the Table 3.2.

Table 3.2: The Gene or QTL Pyramiding in different Crop

Species	Traits	Gene from parent 1	Gene from parent 2	Selection stage	Marker
Barley	Yellow mosaic virus	rym1	rym5	F2	RFLP, CAPS
	Yellow mosaic virus	rym4, rym9	rym4, rym9	F1-derived doubled haploid	RAPD, SSR
	Stripe rust	Rspx	QTL 5	F1-derived doubled haploid	SSR
Rice	Bacterial blight	xa5, xa13	xa4, xa21	F2	RFLP, STS
	Bacterial blight	xa21, Bt	RC7 gene, Bt	F2	STS
	Blast	Pil, Piz-5	Pil, Pita	F2	RFLP, STS
	Brown hopper plant	Bph1	Bph2	F4	STS
	Insect resistance	xa21	Bt	F2	STS
Wheat	Powdery mildew	Pm2	Pm4a	F2	RFLP

Integration of CRISPR-Cas9 for Precise Genome Editing

CRISPR-Cas9 is a revolutionary genome editing technology that allows scientists to precisely modify DNA sequences within an organism's genome. This technology has transformed the field of genetics and has numerous applications in agriculture, medicine, and basic research. The integration of CRISPR-Cas9 for genome editing has provided a powerful tool for targeted modifications in the DNA of various organisms. CRISPR-Cas9 stands for "Clustered Regularly Interspaced Short Palindromic Repeats-CRISPR-associated protein 9." It is a system borrowed from the bacterial immune system that bacteria use to defend against viruses. The key components of CRISPR-Cas9 are

Guide RNA (gRNA)

A synthetic RNA molecule designed to be complementary to the target DNA sequence. The gRNA guides the Cas9 protein to the specific location in the genome where editing is desired.

Cas9 Protein: An enzyme that acts as molecular scissors, capable of cutting both strands of the DNA at the targeted location.

CRISPR-Cas9 is used to develop crops with improved traits such as disease resistance, enhanced nutritional content, and increased yield. CRISPR-Cas9 aids in understanding the functions of genes by knocking them out or modifying their expression. CRISPR-Cas9 offers the potential to correct genetic mutations responsible for diseases.

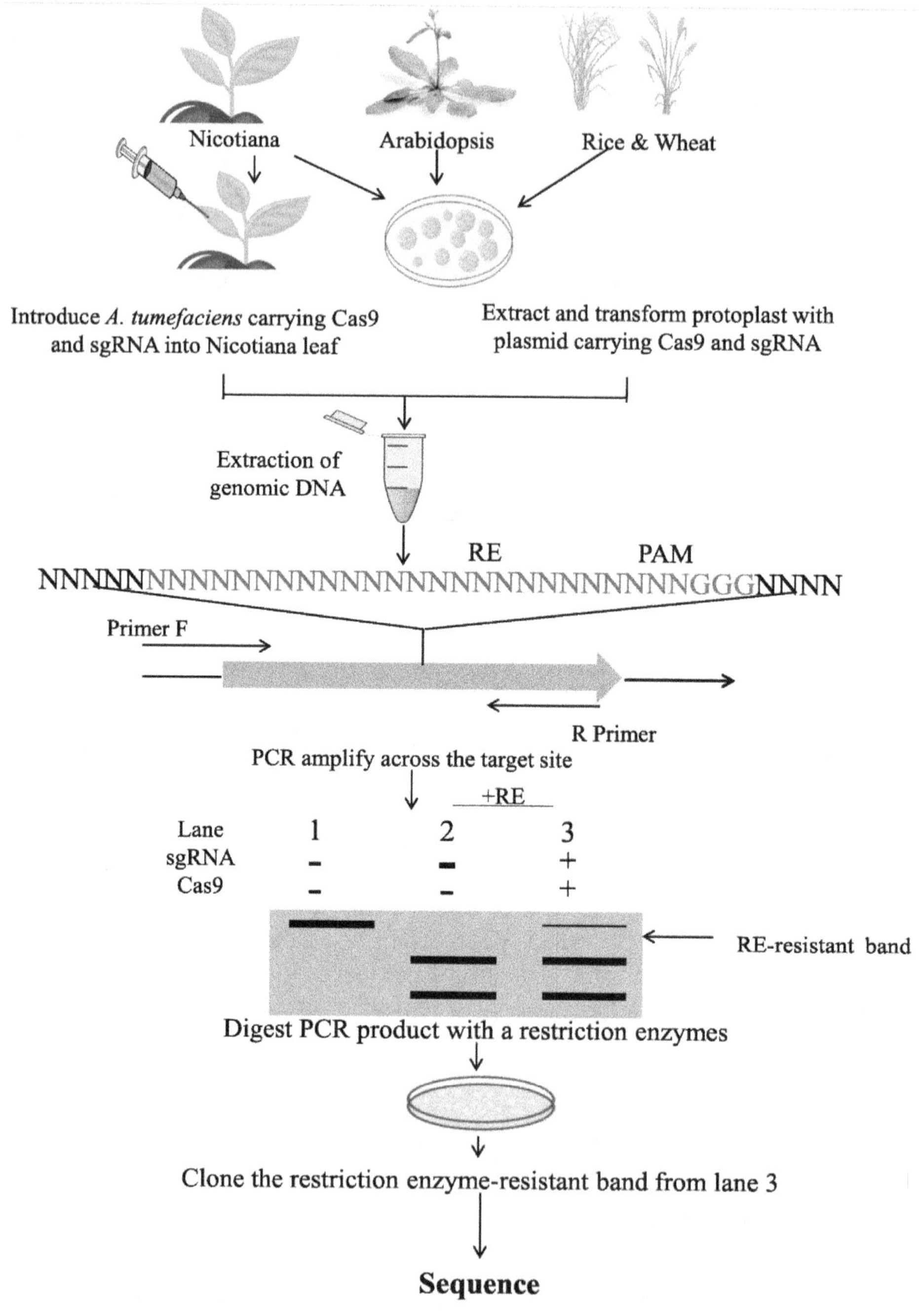

Figure 3.1: Genome Editing for Crop Improvement through the CRISPR/Cas9 Strategy.

Conclusion

In the intricate tapestry of biological research, molecular markers have emerged as indispensable threads, weaving together our understanding of genetics, evolution, and a multitude of applications. As we journey through the realms of crop improvement, conservation genetics, human health, forensics, and phylogenetic studies, the profound impact of molecular markers becomes increasingly evident. Their role in deciphering genetic mysteries and propelling scientific progress is nothing short of transformative, with far-reaching implications for both fundamental knowledge and practical applications.They have illuminated the pathways of evolution, enabling scientists to decipher the intricate connections between species, trace the footsteps of migrations, and reconstruct the branching patterns of the Tree of Life. Through the lens of molecular markers, we have unveiled the genetic secrets behind complex traits, delved into the mechanisms of gene regulation, and uncovered the fingerprints of adaptation imprinted across genomes. Beyond unravelling genetic histories, molecular markers have fuelled the development of innovative solutions that span a spectrum of domains. In agriculture, they have propelled crop improvement to new heights, enabling breeders to harness genetic diversity with precision, resulting in robust and resilient crop varieties. Conservation efforts are guided by the insights gleaned from molecular markers, which identify genetic hotspots and direct us towards tailored strategies for safeguarding biodiversity.

Genome-wide association studies (GWAS) have identified genetic variations linked to diseases, illuminating the genetic basis of illnesses and opening up new treatment possibilities. In the field of forensics, the use of molecular markers has completely changed the way that criminal investigations are conducted by giving investigators concrete ways to positively identify suspects and find the truth even in the most difficult instances.

REFERENCES

Agarwal M, Shrivastava N, Padh H.2008. Advances in molecular marker techniques and their applications in plant sciences. *Plant Cell Rep.*27(4): 617-31.

Bachlava, E., Taylor, C.A., Tang, S. 2012. SNP discovery and development of a high-density genotyping array for sunflower. *PLoS One.*7 (1): e29814.

Baloch FS, Alsaleh A, Shahid MQ.2017. A whole genome DArT seq and SNP analysis for genetic diversity assessment in durum wheat from Central Fertile Crescent. *PloSOne.*12(1): 0167821.

Blears MJ, De Grandis SA, Lee H.1998. Amplified fragment length polymorphism (AFLP): a review of the procedure and its applications. *J Ind Microbiol Biotechnol.*21(3): 99–114.

Chial, H. (2008) DNA fingerprinting using amplified fragment length polymorphisms (AFLP): No genome sequence required. *Nature Education* 1(1): 176

Collard, B.C.Y., and D.J., Mackill. 2007. Marker-assisted selection: an approach for precision plant breeding in the twenty-first century. *Trans. R. Soc.* 363: 557-572.

Jiang GL. 2012.Molecular markers and marker-assisted breeding in plants. In: Andersen SB, editor. Plant breeding from laboratories to fields. *Rijeka: InTech*.p.45–83.

Jones CJ, Edwards KJ, Castaglione S.1997.Reproducibilitytesting of RAPD, AFLP and SSR markers in plants by a network of European laboratories. *Mol Breed*.3(5): 381–390

Mondini L, Noorani A, Pagnotta MA.2009. Assessing plant genetic diversity by molecular tools. Diversity.1(1): 19–35

Morgante M, Hanafey M, Powell W. Microsatellites are preferentially associated with nonrepetitive DNA in plant genomes. *Nat Genet*.2002;30(2): 194–200

Mullis K, Faloona F, Scharf S.1986. Specific enzymatic amplification of DNA in vitro: the polymerase chain reaction. *Cold Spring Harb Symp Quant Biol*.51: 263–273.

Nadeem, M. A., Nawaz, M. A., Shahid, M. Q., Doðan, Y., Comertpay, G., Yýldýz, M.,. and Baloch, F. S. (2017). DNA molecular markers in plant breeding: current status and recent advancements in genomic selection and genome editing. *Biotechnology and Biotechnological Equipment, 32*(2), 261-285.

Williams, JG., Kubelik, AR., Livak KJ.2018.DNA polymorphisms amplified by arbitrary primers are useful as genetic markers. *Nucleic Acids Res*.18(22): 6531–653.

Chapter 4

SNPs and InDels Genotyping: Methods and Application in Plant Breeding

Rashpal Singh[1], Rajvinder Singh[2], Ravindra Kumar Meena[2] and D.N. Kamat[3]*

[1]*Ph.D Scholar, Department of Genetics and Plant Breeding, Maharana Pratap University of Agriculture and Technology, Udaipur – 313 001, Rajasthan*
[2]*Ph.D Scholar, Department of Genetics and Plant Breeding, CCS Haryana Agricultural University, Hisar – 125 004, Haryana*
[3]*Assisstant Scientist, Department of Plant Breeding and Genetics, Dr. Rajendra Prasad Central Agricultural University, Pusa, Samastipur – 848 125, Bihar*
**e-mail: rashpalkhosa9@gmail.com*

ABSTRACT

The most prevalent genetic variation in the plant genome is a single nucleotide polymorphism (SNP). SNPs are crucial markers in numerous researches that connect sequence variants to phenotypic changes; these studies are anticipated to increase our knowledge about crop physiology and reveal the molecular basis of variables that affect yield. The use of single nucleotide polymorphisms (SNPs) as genetic markers for linkage and association studies has received widespread recognition. After SNPs, InDels are the second most common kind of genetic variation. The PCR-amplified InDel polymorphic molecular marker is based on particular primers created from both sides of the location of the sequence of insertion/deletion. The development of the InDel molecular marker, which is located on functional genes, combined with chromosome walking and fine gene mapping, has allowed these molecular markers to be used in the screening of genes related to important economic traits, which is conducive to the further development and utilization of these valuable genes. This chapter provides an overview of genotyping methods for SNPs and InDels and how they are used in plant breeding.

Keywords*: Single nucleotide polymorphisms (SNPs), InDels, Haplotype, Molecular marker.*

INTRODUCTION

The population of 9 billion people will need more food by 2050 than is now being produced by the main crops (Neupane *et al.*, 2022). Climate change, which is expected to worsen in the future with altered rainfall patterns (resulting in floods or drought), extreme weather events, and altering patterns of pathogens and pests in terms of intensity and distribution, makes it even more difficult to feed such a large population (Abberton *et al.*, 2016). In addition to having a significant impact on medicine and public health, Dr. Norman E. Borlaug noted that modern genomics and biotechnological tools have the potential to bring about a second Green Revolution by fusing these priceless scientific methodologies and products with traditional breeding methods (Borlaug, 2000). The creation and use of molecular markers in crop genetics has attracted a lot of attention recently. With the advent of single nucleotide polymorphism (SNP) markers based on next-generation sequencing (NGS) technology, molecular markers have progressed from low-throughput restriction fragment length polymorphisms (RFLPs) to their current state. Single nucleotide polymorphisms, frequently called SNPs, are genomic variants at a single base position in the DNA.Crop genomes include many SNPs, which make them excellent markers for genetic research and molecular breeding. The development of reliable toolkits for use in breeding is also made possible by SNPs from cloned genes and genome-wide linkage and association analyses employing array-based technologies in combination with genotyping by sequencing (GBS).

Crop improvement programs still mainly rely on conventional gel-based markers or other low-throughput platforms for molecular diagnostics, which make it difficult to apply large-scale marker-assisted selection because it takes more time and money. Although advances in gene isolation have produced diagnostic markers for use in breeding, high-throughput tests are still reluctant to adopt these markers in many crops.Several multiplex and single-plex marker platforms with high throughput in rice (Masouleh *et al.*, 2009), wheat (Berard *et al.*, 2009, Bernardo *et al.*, 2015 and Rasheed *et al.*, 2016), legumes (Varshney, 2016) and other crops have been proposed for marker-assisted selection (MAS). However, the deployment of such platforms in crop breeding is constrained by less effective customization, higher costs, desired flexibility and expensive application-related equipment. Although there is still a long way to go, the ultimate goal is to create genotyping systems with ultra-high throughput and low cost for actual breeding. For practical breeding applications, we defined such systems that could quickly genotype tens of thousands of breeding accessions with tens to hundreds of markers.

For a number of crops, SNP chips are presently accessible; however, one drawback is that these readily accessible chips are based on SNPs found in certain genotypes and may not be suited for projects using unrelated genotypes. This requires the development of several chips or the use of technologies that allow for flexible design but are cost-effective (Rasheed, *et al.*, 2017). Single nucleotide polymorphisms (SNPs) have gained universal acceptance as genetic markers for linkage and association research. Because of this, a lot of work has been put into developing precise, quick, and affordable technology for SNP analysis during the past few years, leading to the development of numerous unique methodologies.

SNPs are the most frequent types of DNA variation found in most organisms (Ganal *et al.*, 2009, Hu *et al.*, 2014 and McCouch *et al.*, 2010) but InDels, which are polymorphisms in the length of the DNA sequence caused by the insertion or deletion of one or more nucleotides at a specific site in the genome (Weber *et al.*, 1989), have received a lesser amount of attention than SNPs (Liu *et al.*, 2013). However, in humans and plants, InDels are the second most prevalent kind of genetic variation (Pena and Pena 2012, Song *et al.*, 2015 and Vishwakarma *et al.*, 2017) and they benefited from codominant inheritance, a multiallelic nature, and wide genome coverage (Das *et al.*, 2015). InDels may be quickly detected in comparison to SNPs based on their size (Song *et al.*, 2015). InDels have indeed developed into potent molecular markers for species diagnostics (Yamaki *et al.*, 2013), evolutionary studies (Weber *et al.*, 2002), genetic linkage map construction (Song *et al.*, 2015 and Li *et al.*, 2015) and marker-assisted selection (MAS) breeding (Liu *et al.*, 2013).

Types of SNPs

Single-nucleotide polymorphisms (SNPs, pronounced as "snips") describe variation amongdifferent individuals of a species for single basepairs at thecorresponding sites of their genomes. An SNP locus is a place in the genome where different nucleotides can be found in the same DNA strand of different members of the same species. The sequence that surrounds the polymorphic nucleotide must thus serve as the basis of each SNP locus. There are four possible alleles at an SNP locus, each of which is represented by one of the four DNA nucleotides (Figure). SNPs are often rated as biallelic markers in any scenario. SNPs are created either by transversion (A or G to C or T and vice versa) or transition (C to T, A to G, and vice versa). Transversions appear to be less common overall than transitions (Figure 4.1).

SNPs can be categorized in a number of ways depending on several factors, such as genomic position and phenotypic impact. Noncoding SNPs (ncSNPs) are SNPs that are found in non-coding sections of the genome; these SNPs are sometimes referred to as intronic SNPs when they are found in introns. Similar

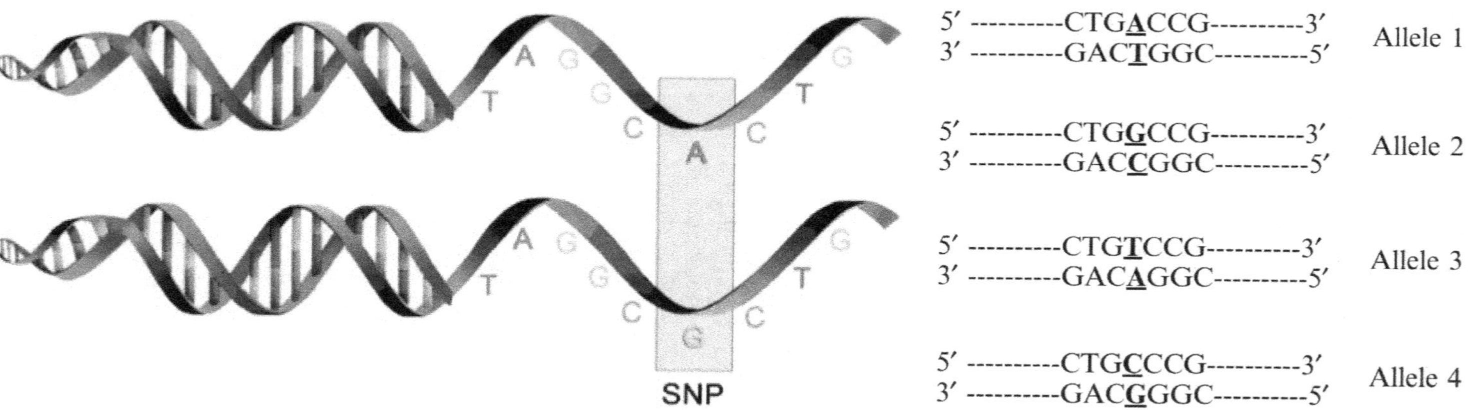

Figure 4.1: Four Possible Alleles at SNP Locus.

to copy SNPs (also known as cSNPs or cDNA SNPs: SNPs found in cDNAs), exonic SNPs, also known as coding SNPs, are found in exons. A synonymous SNP (synSNP) is an exonic SNP that does not affect the amino acid sequence of the affected protein, whereas a nonsynonymous SNP (nsSNP) changes the amino acid sequence. GenicSNPs occur in genes and would include intronic and exonic SNPs. SNPs located in the promoter region of the concerned gene are called promoter SNPs or pSNPs and are included in genic SNPs. Functional SNPs or candidate SNPs are genic SNPs that have the potential to alter the way of gene functions and lead to phenotypic changes. However, most SNPs fall into the category of anonymous SNPs, which do not alter a gene's function or phenotype. In silico SNPs (isSNPs) or electronic SNPs (eSNPs) are the most common terms for SNPs found by mining ESTs or genomic databases; these "virtual" polymorphisms need to be confirmed by resequencing.Simple SNPs and hemi-SNPs or homoeo-SNPs are the different types of SNPs found in polyploid organisms. A simple or true SNP does not detect variations in the other genome(s) of a polyploid species; rather, it detects allelic variation across homologous loci of the same genome that are present in the same or other polyploid species. However, hemi-SNPs or homoeo-SNPs can find homologous or paralogous sites in two or more genomes of polyploid species and the genomes of their diploid ancestors. The value of these hemi-SNPs for mapping is restricted (Deschamps and Campbell 2010), while true SNPs would show typical diploid segregation in most mapping populations, is quite frequent (10-30 per cent), and would be the most useful for mapping.

SNPs Discovery in Complex Plant Genome

Complex genomes present significant challenges for researchers interested in generating SNPs, but SNP discovery in crops with simple genomes is a comparatively easy procedure. The extremely repetitive character of plant genomes is one of the main issues (Meyers *et al.*, 2001). In order to avoid repeated regions of the genome, researchers used to rely on various experimental procedures before the development of next-generation sequencing (NGS) technology. These include the experimental identification of SNPs through Sanger sequencing of amplicons produced from unigenes (Wright *et al.*, 2005) and EST database SNP mining for *in silico* SNP identification, followed by PCR-based validation (Batley *et al.*, 2003). The SNPs discovered by SNP mining are often termed in silico SNPs (isSNPs) or electronic SNPs (eSNPs).

Numerous crops have mosaics of dispersed duplicated regions and are ancient tetraploids (Pratap *et al.*, 2012), numerous non-allelic SNPs representing paralogous sequences that were unsuitable for use in molecular breeding were found through *in silico* and experimental mining of EST databases (Choi *et al.*, 2007). Issues related to poor throughput and the high cost of SNP discovery have been removed by the recent development of NGS technologies including 454 Life Sciences (Roche Applied Science, Indianapolis, IN), HiSeq (Illumina, San

Diego, CA), SOLiD, and Ion Torrent (Life Technologies Corporation, Carlsbad, CA) (Dwiningsih *et al.*, 2020).

NGS technology for transcriptome resequencing enables quick and affordable SNP detection inside genes while avoiding highly repetitive portions of a genome (Morozova and Marra, 2008). For the analysis of several plant genomes, including maize (Barbazuk *et al.*, 2007), canola (Trick *et al.*, 2009), eucalyptus (Novaes *et al.*, 2008), sugarcane (Bundock *et al.*, 2009), tree species (Parchman *et al.*, 2010), wheat (Lai *et al.*, 2012), avocado (Kuhn, 2012), and black currant (Russell *et al.*, 2011) this technology was successfully used. The availability of SNPs inside coding sequences is undoubtedly a highly effective tool for molecular geneticists to find a causal mutation (Varshney, 2009). But frequently, QTLs are found in non-coding regulatory sequences like enhancers or locus control regions, which may be positioned several megabases distant from genes inside intergenic gaps (Dean, 2006). It is difficult to find SNPs inside such regulatory regions using transcriptome or exon sequencing. It is crucial to use genome complexity reduction strategies in conjunction with NGS technology in order to find SNPs in a genome-wide manner while avoiding repetitive and duplicated DNA.

Over time, several methods for reducing genomic complexity have been developed, including high Cot selection, methylation filtering, and microarray-based genomic selection. These methods mostly decrease repetitive sequences, but they are unable to identify and remove duplicate sequences, which results in the identification of false-positive SNPs. Recent methods for reducing genomic complexity include Both the Restriction Site Associated DNA (RAD) (Floragenics, Eugene, OR, USA) and Complexity Reduction of Polymorphic Sequences (CRoPS) (Keygene N.V., Wageningen, The Netherlands) are computationally powerful tools that can filter out duplicated SNPs. SNPs have been discovered in crops with and without reference genome sequences using these methods.

The NimbleGen sequence capture technique (Roche Applied Science, IN) was first created for human disease diagnostic research (Hodges *et al.*, 2007), and it has increased the throughput and coverage level of gene-based SNP identification in plants (Springer *et al.*, 2009). This method involves microarray-based exon sequence capture and enrichment, followed by next-generation sequencing (NGS) for targeted resequencing.

SNPs Validation and Genotyping

It is not always possible to turn a found SNP into a reliable marker, even with the aid of sophisticated tools and access to reference sequences. It is necessary to validate the found SNP to make sure that it is a Mendelian locus and not a product of sequencing error. Resequencing the relevant genomic area of the selected individuals or lines is one method for SNP validation, which verifies that the founded SNPs indeed correspond to polymorphisms. A segregating population is more informative as a validation panel than a group of unrelated lines because it enables examination of the discriminatory power and segregation patterns of a marker, assisting the researcher in determining whether it is a Mendelian locus or a duplicated/repetitive sequence that eluded the software filter (Mammadov *et al.*, 2012).

The many SNP genotyping techniques/platforms can score a single SNP marker or very many markers utilizing high-density SNP chips, and they can be used for a variety of purposes. These techniques rely on approaches that can discriminate between an oligonucleotide and the template DNA strand that are a perfect match and those that have one base out of place. These tactics rely on DNA replication, single-strand invasion with cleavage of the displaced strand, primer extension, oligonucleotide ligation, nucleic acid hybridization, and other techniques (Sobrino *et al.*, 2005).

The most widely used high throughput screening (HTP) assays/chemistries and genotyping systems at the moment for SNP validation and genotyping are KBiosciences' Competitive Allele Specific PCR (KASPar) coupled with the SNP Line platform (SNP Line XL), Life Technologies' TaqMan assay coupled with Open Array platform, and Illumina's BeadArray-based Golden Gate (GG) and Infinium assays. These contemporary genotyping assays and platforms vary from one another in terms of their chemistry, price, throughput of genotyped samples, the number of validated SNPs, length of the SNP context sequence, the total number of SNPs to genotype, and lastly the amount of funding available to the research project all influence the choice of chemical and genotyping platform. Kumpatla (2012) described comparative analyses of these four genotyping assays.

SNPs: Emerging Molecular Marker

The adoption of molecular breeding of crops has been made possible by the use of molecular markers, which have revolutionized the efficiency and accuracy of plant genetic analysis. In the previous few decades, both the development of marker systems and the corresponding detection platforms have made significant advances. Single nucleotide polymorphism (SNP) markers have rapidly risen to the lead in molecular genetics in recent years due to their prevalence in genomes and suitability for high-throughput detection formats and platforms. Due to the growing amount of sequence data in public databases, computational methods dominate SNP discovery techniques; nonetheless, complex genomes present unique difficulties for the identification of meaningful SNPs, necessitating alternate approaches in particular crops. The availability of several genotyping platforms and chemistries has increased the popularity and effectiveness of SNP usage. SNPs are essential markers in various studies that link sequence differences to phenotypic changes; these investigations are expected to improve our understanding of the genetic basis of factors that influence yield (Kim and Misra, 2007).

Allelic variations within a genome of a single species can be divided into three main categories: segmental insertions/deletions (InDels), single nucleotide polymorphisms (SNPs), and variations in the number of tandem repeats at a given locus (microsatellites or simple sequence repeats, or SSRs). Researchers have been creating and employing genetic tools called molecular markers to identify and monitor these changes in the DNA levels of individuals of a progeny. The throughput, affordability, and repeatability of the detection method have been the main forces driving the evolution of molecular markers (Bernardo, 2008). All molecular markers can be categorized into three main classes based on detection technique and throughput.: (1) high-throughput (HTP) sequence-based markers: SNPs; (2) medium-throughput, PCR-based markers that include random amplification of polymorphic DNA (RAPD), amplified fragment length polymorphism (AFLP), SSRs; (3) low-throughput, hybridization-based markers such as restriction fragment length polymorphisms (RFLPs) (Botstein *et al.*, 1980).

Due to their reproducibility and codominance, RFLPs were the most extensively employed molecular markers in plant molecular genetics in the late 1980s. However, RFLP detection was an expensive, time-consuming, and laborious operation, rendering these markers finally ineffective. Low-throughput RFLP markers were displaced by the development of PCR technology and its use for the quick detection of polymorphisms (Welshand McClelland, 1990) and a new generation of PCR-based markers appeared at the start of the 1990s. The main PCR-based markers that the research community has been using in various plant systems are RAPD, AFLP, and SSR markers. RAPDs have the capacity to simultaneously find polymorphism sites throughout a genome. Due to the non-specific binding of short, random primers, they are, however, anonymous, and their level of reproducibility is quite poor. The lengthier +1 and +3 selective primers and the presence of discriminatory nucleotides at the 3′ end of each primer provide AFLPs with a very high level of repeatability and sensitivity despite the fact that they are also anonymous. Because of this, AFLP markers are still widely used in molecular genetics studies of crops with scant or no reference genome sequences. However, due to the lengthy and difficult detection process, which was not automatable additionally, AFLP markers did not find broad application in molecular breeding. SSRs were able to completely overcome all of the inadequacies of the previously mentioned DNA marker technologies; therefore it was not a surprise that they were dubbed "markers of choice" as soon as they were found in the plant genome. SSRs were no longer anonymous they were also very reproducible, quite polymorphic, and susceptible to automation. SSR markers had dominated all facets of plant molecular genetics and breeding by the late 1990s and the start of the twenty-first century, despite the cost of detection continuing to be high.

SNP markers, however, gradually overthrew the monopoly of medium-throughput SSRs over the past few years. SNPs were first identified in the human genome and have since been found to be the most universal and the most prevalent types of genetic variation among members of the same species. SNPs are biallelic, which makes them less polymorphic than SSR markers, but their abundance, ubiquity, and ability to be automated at high and ultra-high throughputs substantially make up for this drawback (Thomson, 2014). As a result of the next-generation sequencing (NGS) technologies, which offer an extensive range of sequencing information with significant gains in coverage, speed, and cost, the modern genomics landscape of crop plants has undergone a revolution (Sharma *et al.*, 2018). The creation of chip-based marker platforms for ultra-high-throughput genotyping is greatly facilitated by these technologies. Although highly developed genotyping platforms have significantly accelerated genetic mapping and gene discovery research and decreased the time and expense required to genotype large populations, their practical uses in crop improvement are quite restricted.

InDels: Insertion-deletions

Insertion-deletions (InDels) are widely acknowledged as important contributors to the genetic structural changes that can be found throughout the genomes of plants. Some cellular processes, including as replication slippage, transposable elements, and crossing-over, result in InDels (Moghaddam *et al.*, 2014). The InDel process continues to have both positive and negative impacts on distinct genomic locations (Pearson *et al.*, 2005). Due to several advantageous intrinsic genetic attributes, such as co-dominant inheritance and multi-allelic with genome-wide dispersion, InDels are clearly a useful marker system for genetic investigations in crops and a nice supplement to other sequence-based genetic markers (Vishwakarma *et al.*, 2017, Lv *et al.*, 2016)). InDel markers may be selected based on their expected fragment size and confirmed in genetic populations/germplasm using straightforward and affordable agarose gels, in addition to being simple to detect in the genome. InDels is a suitable marker system for numerous translational genomics investigations in agricultural plants because of these characteristics.

Contrary to SNP, InDel markers are typically thought of as breeder-friendly markers with minimal infrastructure needs. Their products can be detected using polyacrylamide gel electrophoresis (PAGE) or straightforward gel-based size separation techniques in conventional genetics and breeding laboratories. InDels markers are also frequently amplified without stutter bands, which increases their value. The stutter bands are caused by enzyme slippage during amplification and are especially prevalent for SSR markers with dinucleotide patterns. As the alleles repeat numbers rise, the problem becomes more severe (Guichoux *et al.*, 2011, Hite *et al.*, 1996).

Several studies also discovered that InDels are more polymorphic than microsatellite markers. (Wu *et al.*, 2014 and Liu, 2013). The genotyping of InDels and microsatellites in the same wolves also demonstrates the strong correlation between the polymorphism levels of the two marker types. The use of InDels as genetic markers in non-model species has been limited up to now by a lack of comprehensive genome sequencing data (Vali *et al.*, 2008). It can be expected that this will come to change in the near future. Insertions-deletions can be used effectively for QTL mapping applications in non-model organisms, and they can be beneficial substitutes for SNP and microsatellite markers, particularly for characterizing the genetic architecture in numerous crosses and extensive pedigrees (Vasemagi *et al.*, 2010).

Sequencing can identify single nucleotide insertions and deletions up to a size somewhat smaller than the PCR amplicon, when there has been an insertion or deletion, two overlapping traces will appear in the chromatogram. Insertions/deletions can also be identified by PCR fragment size analysis. Small insertion/deletion (InDel) mutation genotyping methods currently used are expensive, time-consuming, and occasionally inaccurate. In order to address this, scientists created a technique for small indel genotyping in a single polymerase chain reaction, with the ability to distinguish between wild-type, heterozygous, and mutant alleles by band pattern in standard agarose gel electrophoresis (Lin *et al.*, 2020). Lin and coworkers demonstrate this method with multiple genes to distinguish 10 bp, 4 bp, and even 1 bp deletions from the wild type. Restrictions-site associated DNA sequencing (RAD-seq) was used by Zhu *et al.* (2018) to find SNP and InDel markers for building a high-density SNP linkage map in Vitis. Kizil *et al.* (2020) created InDel markers for 95 sesame cultivars using data from double-digested restriction site-associated DNA sequencing (ddRAD-seq).

Due to their importance in crop genomic studies and as a useful supplement to both SNPs and SSRs markers, InDel markers have been frequently found in rice (Wu *et al.*, 2013), barley (Zhou, *et al.*, 2015), oil rapes (Liu, 2013 and Mahmood *et al.*, 2016), maize (Liu, 2015), and other plants (Moghaddam, 2014, Lv *et al.*, 2013 and Yang *et al.*, 2018).

GINDEL (Genotyping Insertions and Deletions from Sequence Reads)

Several increasingly complex computer methods for identifying indels from sequence reads have been developed during the past few years. Let's say that we have found indels in the genome. The genotypes of these indels are the next obvious research subject. Indel genotype calling approach (called GINDEL) employs machine learning to combine features, which allows it to make better use of the data. Each deletion is handled separately by GINDEL, which can identify the genotypes of deletions in populations that contain both a single individual and multiple individuals. GINDEL can identify insertion genotypes, whereas existing methods can only identify deletion genotypes. In terms of genotyping correctness and efficiency on both simulated data and actual population sequencing data, GINDEL surpasses Genome STRiP, Clever-sv, and Pindel (Chu *et al.*, 2014).

Haplotype; Efficient Crop Breeding Tool

The term "haplotype" refers to a group of alleles for various polymorphisms (such as SNPs, insertions/deletions (InDels), and other markers or variants) that are present on the same chromosome and are inherited together with the least amount of contemporary recombination (Stram, 2017 and Garg, 2021). In other words, a set of close genomic structural changes, such as polymorphic SNPs, with significant linkage disequilibrium (LD) between them is referred to as a haplotype (Maldonado *et al.*, 2019). A given stretch of chromosomal DNA has two haplotypes for each individual, although multiple haplotypes can be found for the same stretch in populations (Ammar *et al.*, 2015). As shown in Figure 4.2, two or more polymorphic SNPs of the haploid sequences inheritedtogether as a unit constitute a haplotype (Stram, 2017). There is evidence that LD-based methods are more effective at defining haplotypes in genomic and chromosomal regions (Qian *et al.*, 2017). Numerous variables, including the method of pollination, population size and structure, mutation rate, genetic drift, frequency of recombination, and the type of selection applied to a particular chromosomal fragment, all affect the level of LD in a population (Gupta *et al.*, 2005).

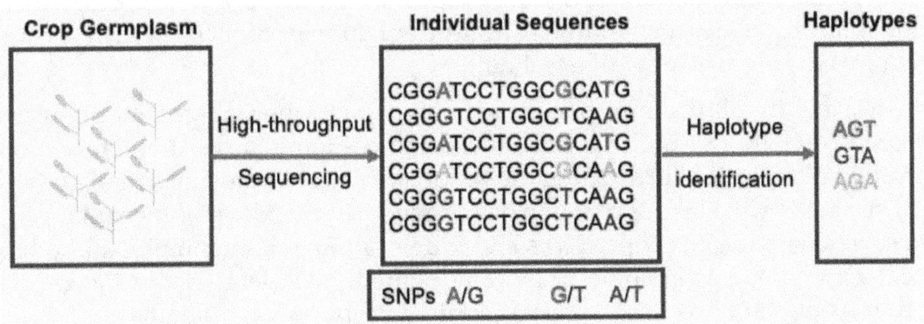

Figure 4.2: Haplotypes from Haploid Crop Genome Sequences.

Recent genome-wide association (GWA) research using empirical and simulated data have shown that haplotype blocks have superior mapping accuracy and power than individual SNPs for the discovery of QTLs and genes (Liu *et al.*, 2019, Lu *et al.*, 2012, Diaye *et al.*, 2017, Lujan *et al.*, 2019 and Srivastava *et al.*, 2020). In contrast to SNP markers, which are biallelic in nature, Stephens *et al.* (2001) showed that haplotype blocks are more informative due to their multi-allelic character.

Generally, genotyping only a small number ofcarefully selected SNP loci from a haplotypeblock allows the deduction of genotypes at theremaining SNP loci of the block; these SNPs aretermed "tag" SNPs. Bhat *et al.* (2021) found that haplotype variations were more abundant than SNPs, indicating that the genes in the haplotype have undergone repeated recombination and mutation processes. Furthermore, compared to individual SNPs, the haplotype-based approach is anticipated to better limit false positives and disclose the intricate mechanism of causative haplotypes. Haplotype-based analysis, in example, can record the interactions between SNPs at a locus that are epistatic (Clark, 2004, Bardel *et al.*, 2005 and Qian *et al.*, 2017).

Conclusion

The availability of SNP markers across the whole genome and their suitability for high- to ultra-high-throughput detection technologies have led to their enormous popularity in plant molecular genetics. SNPs, as opposed to older marker systems, allowed for the creation of saturated or supersaturated genetic maps, allowing for the tracking of the whole genome, precise mapping of regions of interest, quick association of markers with a trait, and quick cloning of relevant genes and QTLs. On the other hand, there are several difficulties that must be resolved or overcome while employing SNPs. For instance, to achieve the same or greater potency than that of earlier-generation molecular markers, the biallelic character of SNPs must be balanced by the discovery and use of a greater number of SNPs. Another difficulty is working with polyploid crops, where the relevant SNPs only make up a small portion of the overall polymorphisms. The generation of a sufficient number of SNPs in such species requires the use of innovative techniques.

InDels display a number of advantageous inherent genetic features shared by SNP and SSR markers, including co-dominance, abundance, and random distribution throughout the genome. The complete molecular study is constrained by the lack of research on the genome-wide development of InDels in several significant crops. SNPs and Indels, along with genomics and other next-generation technologies, have undoubtedly accelerated the speed and progress of plant breeding, despite certain drawbacks or difficulties.

REFERENCES

Abberton, M., Batley, J., Bentley, A., Bryant, J., Cai, H., Cockram, J., Costa de Oliveira, A., Cseke, L.J., Dempewolf, H., De Pace, C. and Edwards, D., 2016. Global agricultural intensification during climate change: a role for genomics. *Plant biotechnology journal*, 14(4), pp.1095-1098.

Ammar, R., Paton, T.A., Torti, D., Shlien, A. and Bader, G.D., 2015. Long read nanopore sequencing for detection of HLA and CYP2D6 variants and haplotypes. *F1000Research, 4.*

Barbazuk, W. Brad, Scott J. Emrich, Hsin D. Chen, Li Li, and Patrick S. Schnable., 2007. "SNP discovery via 454 transcriptome sequencing." *The plant journal* 51, no. 5: 910-918.

Bardel, C., Danjean, V., Hugot, J.P., Darlu, P. and Génin, E., 2005. On the use of haplotype phylogeny to detect disease susceptibility loci. *BMC genetics, 6,* pp.1-13.

Batley, J., Barker, G., O'Sullivan, H., Edwards, K.J. and Edwards, D., 2003. Mining for single nucleotide polymorphisms and insertions/deletions in maize expressed sequence tag data. *Plant physiology*, 132(1), pp.84-91.

Berard, A., Le Paslier, M.C., Dardevet, M., ExbrayatVinson, F., Bonnin, I., Cenci, A., Haudry, A., Brunel, D. and Ravel, C., 2009. Highthroughput single nucleotide polymorphism genotyping in wheat (Triticum spp.). *Plant Biotechnology Journal*, 7(4), pp.364-374.

Bernardo, A., Wang, S., St. Amand, P. and Bai, G., 2015. Using next generation sequencing for multiplexed trait-linked markers in wheat. *PloS one*, 10(12), p.e0143890.

Bernardo, R., 2008. Molecular markers and selection for complex traits in plants: learning from the last 20 years. *Crop science*, 48(5), pp.1649-1664.

Bhat, J.A., Yu, D., Bohra, A., Ganie, S.A. and Varshney, R.K., 2021. Features and applications of haplotypes in crop breeding. CommunBiol 4: 1–12.

Borlaug, N.E., 2000. Ending world hunger. The promise of biotechnology and the threat of antiscience zealotry. *Plant physiology*, 124(2), pp.487-490.

Botstein, D., White, R.L., Skolnick, M. and Davis, R.W., 1980. Construction of a genetic linkage map in man using restriction fragment length polymorphisms. *American journal of human genetics*, 32(3), p.314.

Bundock, P.C., Eliott, F.G., Ablett, G., Benson, A.D., Casu, R.E., Aitken, K.S. and Henry, R.J., 2009. Targeted single nucleotide polymorphism (SNP) discovery in a highly polyploid plant species using 454 sequencing. *Plant Biotechnology Journal*, 7(4), pp.347-354.

Choi, I.Y., Hyten, D.L., Matukumalli, L.K., Song, Q., Chaky, J.M., Quigley, C.V., Chase, K., Lark, K.G., Reiter, R.S., Yoon, M.S. and Hwang, E.Y., 2007. A soybean transcript map: gene distribution, haplotype and single-nucleotide polymorphism analysis. *Genetics*, 176(1), pp.685-696.

Chu, C., Zhang, J. and Wu, Y., 2014. GINDEL: accurate genotype calling of insertions and deletions from low coverage population sequence reads. *PloS one*, 9(11), p.e113324.

Clark, A.G., 2004. The role of haplotypes in candidate gene studies. *Genetic Epidemiology: The Official Publication of the International Genetic Epidemiology Society*, 27(4), pp.321-333.

Das, S., Upadhyaya, H.D., Srivastava, R., Bajaj, D., Gowda, C.L.L., Sharma, S., Singh, S., Tyagi, A.K. and Parida, S.K., 2015. Genome-wide insertion–deletion (InDel) marker discovery and genotyping for genomics-assisted breeding applications in chickpea. *DNA Research*, 22(5), pp.377-386.

Dean, A., 2006. On a chromosome far, far away: LCRs and gene expression. *TRENDS in Genetics*, 22(1), pp.38-45.

Deschamps, S. and Campbell, M.A., 2010. Utilization of next-generation sequencing platforms in plant genomics and genetic variant discovery. *Molecular breeding, 25*, pp.553-570.

Dwiningsih, Y., Rahmaningsih, M. and Alkahtani, J., 2020. Development of single nucleotide polymorphism (SNP) markers in tropical crops. *Advance Sustainable Science, Engineering and Technology, 2*(2), p.343558.

Ganal, M.W., Altmann, T. and Röder, M.S., 2009. SNP identification in crop plants. *Current opinion in plant biology, 12*(2), pp.211-217.

Garg, S., 2021. Computational methods for chromosome-scale haplotype reconstruction. *Genome biology, 22*(1), pp.1-24.

Guichoux, E., Lagache, L., Wagner, S., Chaumeil, P., Léger, P., Lepais, O., Lepoittevin, C., Malausa, T., Revardel, E., Salin, F. and Petit, R.J., 2011. Current trends in microsatellite genotyping. *Molecular ecology resources, 11*(4), pp.591-611.

Gupta, P.K., Rustgi, S. and Kulwal, P.L., 2005. Linkage disequilibrium and association studies in higher plants: present status and future prospects. *Plant molecular biology, 57*, pp.461-485.

Hite, J.M., Eckert, K.A. and Cheng, K.C., 1996. Factors affecting fidelity of DNA synthesis during PCR amplification of d (CA) n• d (GT) n microsatellite repeats. *Nucleic acids research, 24*(12), pp.2429-2434.

Hodges, E., Xuan, Z., Balija, V., Kramer, M., Molla, M.N., Smith, S.W., Middle, C.M., Rodesch, M.J., Albert, T.J., Hannon, G.J. and McCombie, W.R., 2007. Genome-wide in situ exon capture for selective resequencing. *Nature genetics, 39*(12), pp.1522-1527.

Hu, Y., Mao, B., Peng, Y., Sun, Y., Pan, Y., Xia, Y., Sheng, X., Li, Y., Tang, L., Yuan, L. and Zhao, B., 2014. Deep re-sequencing of a widely used maintainer line of hybrid rice for discovery of DNA polymorphisms and evaluation of genetic diversity. *Molecular Genetics and Genomics, 289*, pp.303-315.

Khan, M.A., Han, Y., Zhao, Y.F., Troggio, M. and Korban, S.S., 2012. A multi-population consensus genetic map reveals inconsistent marker order among maps likely attributed to structural variations in the apple genome. *PloS one, 7*(11), p.e47864.

Kim, S. and Misra, A., 2007. SNP genotyping: technologies and biomedical applications. *Annu. Rev. Biomed. Eng., 9*, pp.289-320.

Kizil, S., Basak, M., Guden, B., Tosun, H.S., Uzun, B. and Yol, E., 2020. Genome-wide discovery of InDel markers in sesame (Sesamum indicum L.) using ddRADSeq. *Plants, 9*(10), p.1262.

Kuhn, D., 2012, January. Design of an Illumina Infinium 6k SNP chip for genotyping two large avocado mapping populations. In *Proceedings of the 20th Conference on Plant and Animal Genome*.

Kumpatla, S.P., 2012. Genomics-assisted plant breeding in the 21st century: technological advances and progress. *Science web academic papers collection*.

Lai, K., Duran, C., Berkman, P.J., Lorenc, M.T., Stiller, J., Manoli, S., Hayden, M.J., Forrest, K.L., Fleury, D., Baumann, U. and Zander, M., 2012. Single nucleotide polymorphism discovery from wheat nextgeneration sequence data. *Plant biotechnology journal, 10*(6), pp.743-749.

Li, W., Cheng, J., Wu, Z., Qin, C., Tan, S., Tang, X., Cui, J., Zhang, L. and Hu, K., 2015. An InDel-based linkage map of hot pepper (Capsicum annuum). *Molecular Breeding, 35*, pp.1-10.

Lin, B., Sun, J. and Fraser, I.D., 2020. Single-tube genotyping for small insertion/deletion mutations: simultaneous identification of wild type, mutant and heterozygous alleles. *Biology Methods and Protocols, 5*(1), p.bpaa007.

Liu, B., Wang, Y., Zhai, W., Deng, J., Wang, H., Cui, Y., Cheng, F., Wang, X. and Wu, J., 2013. Development of InDel markers for Brassica rapa based on whole-genome re-sequencing. *Theoretical and Applied Genetics, 126*, pp.231-239.

Liu, F., Schmidt, R.H., Reif, J.C. and Jiang, Y., 2019. Selecting closely-linked SNPs based on local epistatic effects for haplotype construction improves power of association mapping. *G3: Genes, Genomes, Genetics, 9*(12), pp.4115-4126.

Liu, J., Qu, J., Yang, C., Tang, D., Li, J., Lan, H. and Rong, T., 2015. Development of genome-wide insertion and deletion markers for maize, based on next-generation sequencing data. *BMC genomics, 16*(1), pp.1-9.

Lu, X., Wang, L., Chen, S., He, L., Yang, X., Shi, Y., Cheng, J., Zhang, L., Gu, C.C., Huang, J. and Wu, T., 2012. Genome-wide association study in Han Chinese identifies four new susceptibility loci for coronary artery disease. *Nature genetics, 44*(8), pp.890-894.

Lujan Basile, S.M., Ramírez, I.A., Crescente, J.M., Conde, M.B., Demichelis, M., Abbate, P., Rogers, W.J., Pontaroli, A.C., Helguera, M. and Vanzetti, L.S., 2019. Haplotype block analysis of an Argentinean hexaploid wheat collection and GWAS for yield components and adaptation. *BMC plant biology, 19*, pp.1-16.

Lv, H.H., Yang, L.M., Kang, J.G., Wang, Q.B., Wang, X.W., Fang, Z.Y., Liu, Y.M., Zhuang, M., Zhang, Y.Y., Lin, Y. and Yang, Y.H., 2013. Development of InDel markers linked to Fusarium wilt resistance in cabbage. *Molecular breeding, 32*, pp.961-967.

Lv, Y., Liu, Y. and Zhao, H., 2016. mInDel: a high-throughput and efficient pipeline for genome-wide InDel marker development. *BMC genomics, 17*(1), pp.1-5.

Mahmood, S., Li, Z., Yue, X., Wang, B., Chen, J. and Liu, K., 2016. Development of INDELs markers in oilseed rape (*Brassica napus* L.) using re-sequencing data. *Molecular Breeding, 36*, pp.1-13.

Maldonado, C., Mora, F., Scapim, C.A. and Coan, M., 2019. Genome-wide haplotype-based association analysis of key traits of plant lodging and architecture of maize identifies major determinants for leaf angle: Hap LA4. *PloS one, 14*(3), p.e0212925.

Mammadov, J., Aggarwal, R., Buyyarapu, R. and Kumpatla, S., 2012. SNP markers and their impact on plant breeding. *International journal of plant genomics, 2012*.

Masouleh, A.K., Waters, D.L., Reinke, R.F. and Henry, R.J., 2009. A highthroughput assay for rapid and simultaneous analysis of perfect markers for important quality and agronomic traits in rice using multiplexed MALDITOF mass spectrometry. *Plant biotechnology journal, 7*(4), pp.355-363.

McCouch, S.R., Teytelman, L., Xu, Y., Lobos, K.B., Clare, K., Walton, M., Fu, B., Maghirang, R., Li, Z., Xing, Y. and Zhang, Q., 2002. Development and mapping of 2240 new SSR markers for rice (Oryza sativa L.). *DNA research, 9*(6), pp.199-207.

McCouch, S.R., Zhao, K., Wright, M., Tung, C.W., Ebana, K., Thomson, M., Reynolds, A., Wang, D., DeClerck, G., Ali, M.L. and McClung, A., 2010. Development of genome-wide SNP assays for rice. *Breeding Science, 60*(5), pp.524-535.

Meyers, B.C., Tingey, S.V. and Morgante, M., 2001. Abundance, distribution, and transcriptional activity of repetitive elements in the maize genome. *Genome Research, 11*(10), pp.1660-1676.

Moghaddam, S.M., Song, Q., Mamidi, S., Schmutz, J., Lee, R., Cregan, P., Osorno, J.M. and McClean, P.E., 2014. Developing market class specific InDel markers from next generation sequence data in Phaseolus vulgaris L. *Frontiers in Plant Science, 5*, p.185

Morozova, O. and Marra, M.A., 2008. Applications of next-generation sequencing technologies in functional genomics. *Genomics, 92*(5), pp.255-264.

N'Diaye, A., Haile, J.K., Cory, A.T., Clarke, F.R., Clarke, J.M., Knox, R.E. and Pozniak, C.J., 2017. Single marker and haplotype-based association analysis of semolina and pasta colour in elite durum wheat breeding lines using a high-density consensus map. *PLoS One, 12*(1), p.e0170941.

Neupane, D., Adhikari, P., Bhattarai, D., Rana, B., Ahmed, Z., Sharma, U. and Adhikari, D., 2022. Does climate change affect the yield of the top three cereals and food security in the world?. *Earth, 3*(1), pp.45-71.

Novaes, E., Drost, D.R., Farmerie, W.G., Pappas, G.J., Grattapaglia, D., Sederoff, R.R. and Kirst, M., 2008. High-throughput gene and SNP discovery in

Eucalyptus grandis, an uncharacterized genome. *BMC genomics*, *9*(1), pp.1-14.

Parchman, T.L., Geist, K.S., Grahnen, J.A., Benkman, C.W. and Buerkle, C.A., 2010. Transcriptome sequencing in an ecologically important tree species: assembly, annotation, and marker discovery. *BMC genomics*, *11*(1), pp.1-16.

Pearson, C.E., Edamura, K.N. and Cleary, J.D., 2005. Repeat instability: mechanisms of dynamic mutations. *Nature Reviews Genetics*, *6*(10), pp.729-742.

Pena, H.B. and Pena, S.D., 2012. Automated genotyping of a highly informative panel of 40 short insertion-deletion polymorphisms resolved in polyacrylamide gels for forensic identification and kinship analysis. *Transfusion Medicine and Hemotherapy*, *39*(3), pp.211-216.

Pratap, A., Gupta, S., Kumar, J. and Solanki, R., 2012. "Soybean," *Technological Innovations in Major World Oil Crops*, vol. 1, pp. 293–321.

Qian, L., Hickey, L.T., Stahl, A., Werner, C.R., Hayes, B., Snowdon, R.J. and Voss-Fels, K.P., 2017. Exploring and harnessing haplotype diversity to improve yield stability in crops. *Frontiers in plant science*, *8*, p.1534.

Rasheed, A., Hao, Y., Xia, X., Khan, A., Xu, Y., Varshney, R.K. and He, Z., 2017. Crop breeding chips and genotyping platforms: progress, challenges, and perspectives. *Molecular plant*, *10*(8), pp.1047-1064.

Rasheed, A., Wen, W., Gao, F., Zhai, S., Jin, H., Liu, J., Guo, Q., Zhang, Y., Dreisigacker, S., Xia, X. and He, Z., 2016. Development and validation of KASP assays for genes underpinning key economic traits in bread wheat. *Theoretical and Applied Genetics*, *129*, pp.1843-1860.

Russell, J.R., Bayer, M., Booth, C., Cardle, L., Hackett, C.A., Hedley, P.E., Jorgensen, L., Morris, J.A. and Brennan, R.M., 2011. Identification, utilisation and mapping of novel transcriptome-based markers from blackcurrant (Ribes nigrum). *BMC Plant Biology*, *11*(1), pp.1-11.

Sharma, S.K., Bolser, D., de Boer, J., Sønderkær, M., Amoros, W., Carboni, M.F., D'Ambrosio, J.M., de la Cruz, G., Di Genova, A., Douches, D.S. and Eguiluz, M., 2013. Construction of reference chromosome-scale pseudomolecules for potato: integrating the potato genome with genetic and physical maps. *G3: Genes, Genomes, Genetics*, *3*(11), pp.2031-2047.

Sharma, T.R., Devanna, B.N., Kiran, K., Singh, P.K., Arora, K., Jain, P., Tiwari, I.M., Dubey, H., Saklani, B.K., Kumari, M. and Singh, J., 2018. Status and prospects of next-generation sequencing technologies in crop plants. *Current Issues in Molecular Biology*, *27*(1), pp.1-36.

Smith, J.S.C., Chin, E.C.L., Shu, H., Smith, O.S., Wall, S.J., Senior, M.L., Mitchell, S.E., Kresovich, S. and Ziegle, J., 1997. An evaluation of the utility of SSR loci as molecular markers in maize (Zea mays L.): comparisons with data from RFLPs and pedigree. *Theoretical and Applied Genetics*, 95, pp.163-173.

Sobrino, B., Brión, M. and Carracedo, A., 2005. SNPs in forensic genetics: a review on SNP typing methodologies. *Forensic science international*, 154(2-3), pp.181-194.

Somers, D.J., Isaac, P. and Edwards, K., 2004. A high-density microsatellite consensus map for bread wheat (Triticum aestivum L.). *Theoretical and applied genetics*, 109, pp.1105-1114.

Song, X., Wei, H., Cheng, W., Yang, S., Zhao, Y., Li, X., Luo, D., Zhang, H. and Feng, X., 2015. Development of INDEL markers for genetic mapping based on whole genome resequencing in soybean. *G3: Genes, Genomes, Genetics*, 5(12), pp.2793-2799.

Springer, N.M., Ying, K., Fu, Y., Ji, T., Yeh, C.T., Jia, Y., Wu, W., Richmond, T., Kitzman, J., Rosenbaum, H. and Iniguez, A.L., 2009. Maize inbreds exhibit high levels of copy number variation (CNV) and presence/absence variation (PAV) in genome content. *PLoS genetics*, 5(11), p.e1000734.

Srivastava, R.K., Singh, R.B., Pujarula, V.L., Bollam, S., Pusuluri, M., Chellapilla, T.S., Yadav, R.S. and Gupta, R., 2020. Genome-wide association studies and genomic selection in pearl millet: Advances and prospects. *Frontiers in Genetics*, 10, p.1389.

Stephens, M., Smith, N.J. and Donnelly, P., 2001. A new statistical method for haplotype reconstruction from population data. *The American Journal of Human Genetics*, 68(4), pp.978-989.

Stram, D.O., 2017. Multi-SNP haplotype analysis methods for association analysis. *Statistical Human Genetics: Methods and Protocols*, pp.485-504.

Thomson, M.J., 2014. High-throughput SNP genotyping to accelerate crop improvement. *Plant Breeding and Biotechnology*, 2(3), pp.195-212.

Trick, M., Long, Y., Meng, J. and Bancroft, I., 2009. Single nucleotide polymorphism (SNP) discovery in the polyploid Brassica napus using Solexa transcriptome sequencing. *Plant biotechnology journal*, 7(4), pp.334-346.

Vali, U., Brandstrom, M., Johansson, M. and Ellegren, H., 2008. Insertion-deletion polymorphisms (indels) as genetic markers in natural populations. *BMC genetics*, 9(1), pp.1-8.

Varshney, R.K., 2009. Gene-based marker systems in plants: high throughput approaches for marker discovery and genotyping. *Molecular Techniques in Crop Improvement: 2nd Edition*, pp.119-142.

Varshney, R.K., 2016. Exciting journey of 10 years from genomes to fields and markets: some success stories of genomics-assisted breeding in chickpea, pigeonpea and groundnut. *Plant Science, 242,* pp.98-107.

Varshney, R.K., Graner, A. and Sorrells, M.E., 2005. Genic microsatellite markers in plants: features and applications. *TRENDS in Biotechnology, 23*(1), pp.48-55.

Varshney, R.K., Marcel, T.C., Ramsay, L., Russell, J., Röder, M.S., Stein, N., Waugh, R., Langridge, P., Niks, R.E. and Graner, A., 2007. A high density barley microsatellite consensus map with 775 SSR loci. *theoretical and Applied Genetics, 114,* pp.1091-1103.

Vasemagi, A., Gross, R., Palm, D., Paaver, T. and Primmer, C.R., 2010. Discovery and application of insertion-deletion (INDEL) polymorphisms for QTL mapping of early life-history traits in Atlantic salmon. *BMC genomics, 11*(1), pp.1-11.

Vishwakarma, M.K., Kale, S.M., Sriswathi, M., Naresh, T., Shasidhar, Y., Garg, V., Pandey, M.K. and Varshney, R.K., 2017. Genome-wide discovery and deployment of insertions and deletions markers provided greater insights on species, genomes, and sections relationships in the genus Arachis. *Frontiers in Plant Science, 8,* p.2064.

Weber, J.L. and May, P.E., 1989. Abundant class of human DNA polymorphisms which can be typed using the polymerase chain reaction. *American journal of human genetics, 44*(3), p.388.

Weber, J.L., David, D., Heil, J., Fan, Y., Zhao, C. and Marth, G., 2002. Human diallelic insertion/deletion polymorphisms. *The American Journal of Human Genetics, 71*(4), pp.854-862.

Welsh, J. and McClelland, M., 1990. Fingerprinting genomes using PCR with arbitrary primers. *Nucleic acids research, 18*(24), pp.7213-7218.

Wright, S.I., Bi, I.V., Schroeder, S.G., Yamasaki, M., Doebley, J.F., McMullen, M.D. and Gaut, B.S., 2005. The effects of artificial selection on the maize genome. *Science, 308*(5726), pp.1310-1314.

Wu, D.H., Wu, H.P., Wang, C.S., Tseng, H.Y. and Hwu, K.K., 2013. Genome-wide InDel marker system for application in rice breeding and mapping studies. *Euphytica, 192,* pp.131-143.

Wu, J., Li, L.T., Li, M., Khan, M.A., Li, X.G., Chen, H., Yin, H. and Zhang, S.L., 2014. High-density genetic linkage map construction and identification of fruit-related QTLs in pear using SNP and SSR markers. *Journal of experimental botany, 65*(20), pp.5771-5781.

Wu, K., Yang, M., Liu, H., Tao, Y., Mei, J. and Zhao, Y., 2014. Genetic analysis and molecular characterization of Chinese sesame (Sesamum indicum L.) cultivars using Insertion-Deletion (InDel) and Simple Sequence Repeat (SSR) markers. *BMC genetics, 15*(1), pp.1-15.

Xu, Y. and Crouch, J.H., 2008. Markerassisted selection in plant breeding: From publications to practice. *Crop science*, *48*(2), pp.391-407.

Yamaki, S., Ohyanagi, H., Yamasaki, M., Eiguchi, M., Miyabayashi, T., Kubo, T., Kurata, N. and Nonomura, K.I., 2013. Development of INDEL markers to discriminate all genome types rapidly in the genus Oryza. *Breeding science*, *63*(3), pp.246-254.

Yang, Z., Dai, Z., Xie, D., Chen, J., Tang, Q., Cheng, C., Xu, Y., Wang, T. and Su, J., 2018. Development of an InDel polymorphism database for jute via comparative transcriptome analysis. *Genome*, *61*(5), pp.323-327.

Zhou, G., Zhang, Q., Tan, C., Zhang, X.Q. and Li, C., 2015. Development of genome-wide InDel markers and their integration with SSR, DArT and SNP markers in single barley map. *BMC genomics*, *16*(1), pp.1-8.

Zhu, J., Guo, Y., Su, K., Liu, Z., Ren, Z., Li, K. and Guo, X., 2018. Construction of a highly saturated genetic map for Vitis by next-generation restriction site-associated DNA sequencing. *BMC plant biology*, *18*, pp.1-12.

Chapter 5

Transcriptomics: Techniques and Prospects

Ellandula Anvesh[1], Sree Vathsa Sagar U.S.[1], Revanna Swamy K.M.[1] and Srinivasu P.[2]*

[1]*Ph.D Scholor, Department of Genetics and Plant Breeding, TNAU, Coimbatore*
[2]*Ph.D Scholor, Department of Plantation, Spices, Medicinal and Aromatic Crops, HC and RI, TNAU, Coimbatore*
**e-mail: pujaman@gmail.com*

INTRODUCTION

Transcriptomics is a field of molecular biology that focuses on studying the transcriptome of an organism. The transcriptome refers to the complete set of all RNA molecules in a cell or a population of cells. RNA molecules play a crucial role in the cell's functioning by carrying genetic information from DNA to guide protein synthesis. Transcriptomics involves the assessing of patterns of gene expression, including the identification and quantification of different types of RNA molecules, such as messenger RNA (mRNA), transfer RNA (tRNA), and ribosomal RNA (rRNA). The main goals of transcriptomics are to understand how genes are expressed under different conditions, how they interact with each other, how they contribute to various biological process.

Transcriptomics techniques involve the high-throughput analysis of RNA molecules, allowing researchers to examine the entire set of genes being expressed in a given sample. This can be done through methods like microarrays and RNA sequencing (RNA-seq). Microarrays are used to measure the expression levels of a predefined set of genes, while RNA-seq provides a comprehensive view of the transcriptome by sequencing all the RNA molecules present in a sample. Transcriptomics has revolutionized our understanding of gene regulation and expression, providing a powerful tool for studying complex biological systems and their responses to various stimuli. It has broad applications in genetics, molecular biology, medicine, agriculture, and environmental science, contributing to advancements in diverse fields.

Techniques in Transcriptomics

1. Micro-array

DNA microarray technology, especially the utilization of GeneChip microarrays, has become a standardized method for simultaneous gene expression analysis. Recent enhancements in GeneChip microarrays have facilitated comprehensive genome-wide expression analysis, opening up new avenues for investigating the composition, dynamics, and regulation of the transcriptome in plants.

DNA microarrays employ a collection of DNA probes ranging from hundreds to hundreds of thousands, arranged on a solid substrate, to assess the abundance and/or binding capability of DNA or RNA target molecules. These DNA probes utilized in a DNA microarray can encompass amplified cDNA fragments or chemically synthesized DNA oligonucleotides with sequences that correspond to the target sequences. Consequently, DNA microarrays are classified into two types: cDNA microarrays and DNA oligonucleotide probe microarrays. These DNA oligomers are either directly synthesized on the microarray's surface or mechanically deposited onto it and subsequently covalently attached. The samples under scrutiny are fluorescently or radioactively labeled before detection. Upon application of the samples to the DNA microarray, DNA probes bind to nucleic-acid target molecules based on sequence complementarity. The intensity of the fluorescent or radioactive signal emitted by the captured targets correlates with the abundanc target molecules and/or the compatibility of binding between the probe and target molecules. This fluorescence or radioactive signal can then be quantitatively recorded using a charge-coupled device (CCD) or laser scanner. As a result of hundreds to thousands of DNA probes being arrayed on a single microarray, it becomes feasible to analyze numerous different targets concurrently, greatly enhancing the efficiency of this detection methodology.

2. RNA-seq

RNA-seq is a high-throughput sequencing of cDNA, and is based on the direct sequencing of transcripts. This technique is more dynamic, reproducible, and provides a better estimate of the absolute expression levels (Nagalakshmi *et al.*, 2008; Fu *et al.*, 2009). These are the main advantages of RNA-seq compared with microarrays. Another advantage is that the analysis of RNA-seq allows us to identify isoforms of a gene, which are not easily detected using microarrays (Wang *et al.*, 2009).

RNA-seq uses next generation sequencing (NGS) methods to sequence cDNA from RNA of biological samples, producing millions of short reads, whose sizes depend on the platform used (Shendure and Ji, 2008). Usually, reads are mapped against a reference genome and the no. of reads mapped to a region of interest is used to measure the relative abundance of its expression. Moreover, the programs that perform assembly using the reference genome can identify isoforms from transcripts (Oshlack *et al.*, 2010). However, RNA-seq also allows assembly of transcripts without the use of a reference genome (de novo assembly), which makes this technique more advantageous compared with other methods for large-scale analysis of gene expression (Grabherr *et al.*, 2011). After the assembly of the transcripts, the next step is the normalization of the data. An example of a normalization method is performed by Cufflinks, which uses data processed in reads per kilobase of exon model per million mapped reads (RPKM) or fragments per kilobase per million mapped reads (FPKM), when analyzing paired-end sequencing data (Trapnell *et al.*, 2010). These transformations normalize the counts with different library sizes and lengths of the transcripts; a long transcript is expected to produce more reads than a short transcript with the same expression level (Soneson and Delorenzi, 2013). However, other programs use different normalization methods.

The analyses of gene expression using RNA-seq are relatively recent and there is still no consensus on the best methods of assembly, normalization, and statistics to calculate thedifferential expression. Existing methods are still being optimized simultaneously with the development of new methods. Thus, the most appropriate methods of analysis should be carefully selected taking into account each experimental condition. All high-throughput transcriptome analyzes, such as EST libraries, microarrays, and RNA-seq, generate a list with hundreds or even thousands of differentially expressed genes. Meaningful biological interpretation from a list of differentially expressed genes is a major challenge in the study of the transcriptome. To address this question, several bioinformatics tools have been developed. Among these tools are the functional classification of genes based on Gene Ontology (Harris *et al.*, 2004), and gene set enrichment analysis (GSEA), which relates the terms of GO to identify the most representative ontologies in the list of differentially expressed genes (Alexa *et al.*, 2006). Furthermore, other tools have been developed to integrate expression of gene data onto gene regulatory networks or biological pathways, providing a visual representation of the graph and integrated data, such as Cytoscape (Shannon *et al.*, 2003), MapMan (Thimm *et al.*, 2004), GENEVESTIGATOR (Zimmermann *et al.*, 2004) and KaPPA (Tokimatsu *et al.*, 2005).

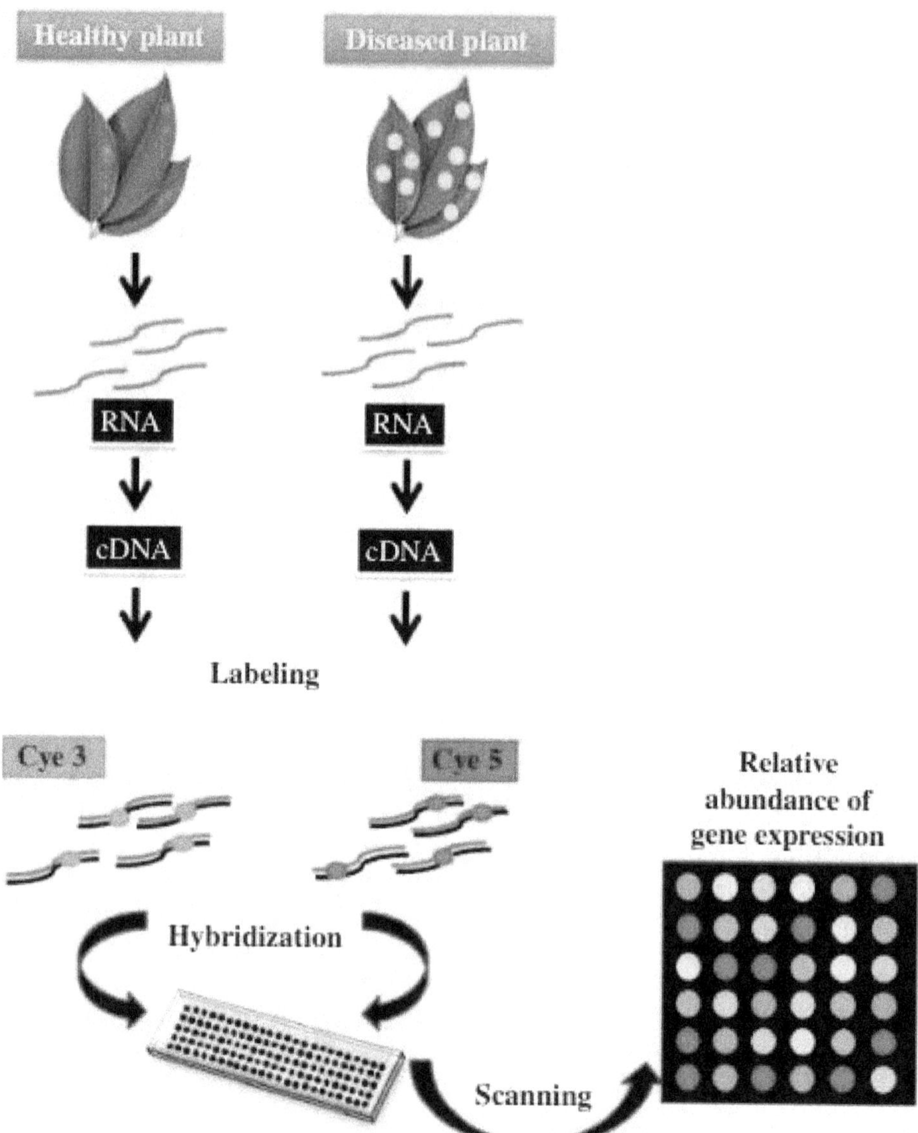

Figure 5.1: Analysis of Gene Expression using a DNA Microarray.

Total RNA is extracted from 2 different samples (healthy and diseased plant), purified, and used as templates for the synthesis of cDNAs, which are labeled with different fluorescent (Cye 3 or Cye 5) dyes for each condition. The labeled cDNAs are mixed and hybridized against the probes (DNA/cDNA known) immobilized on the microarray chip. Laser excitation of the fluorophores produces an emission with a specific spectrum, which is captured by a scanner and analyzed by software. (Borem, 2014)

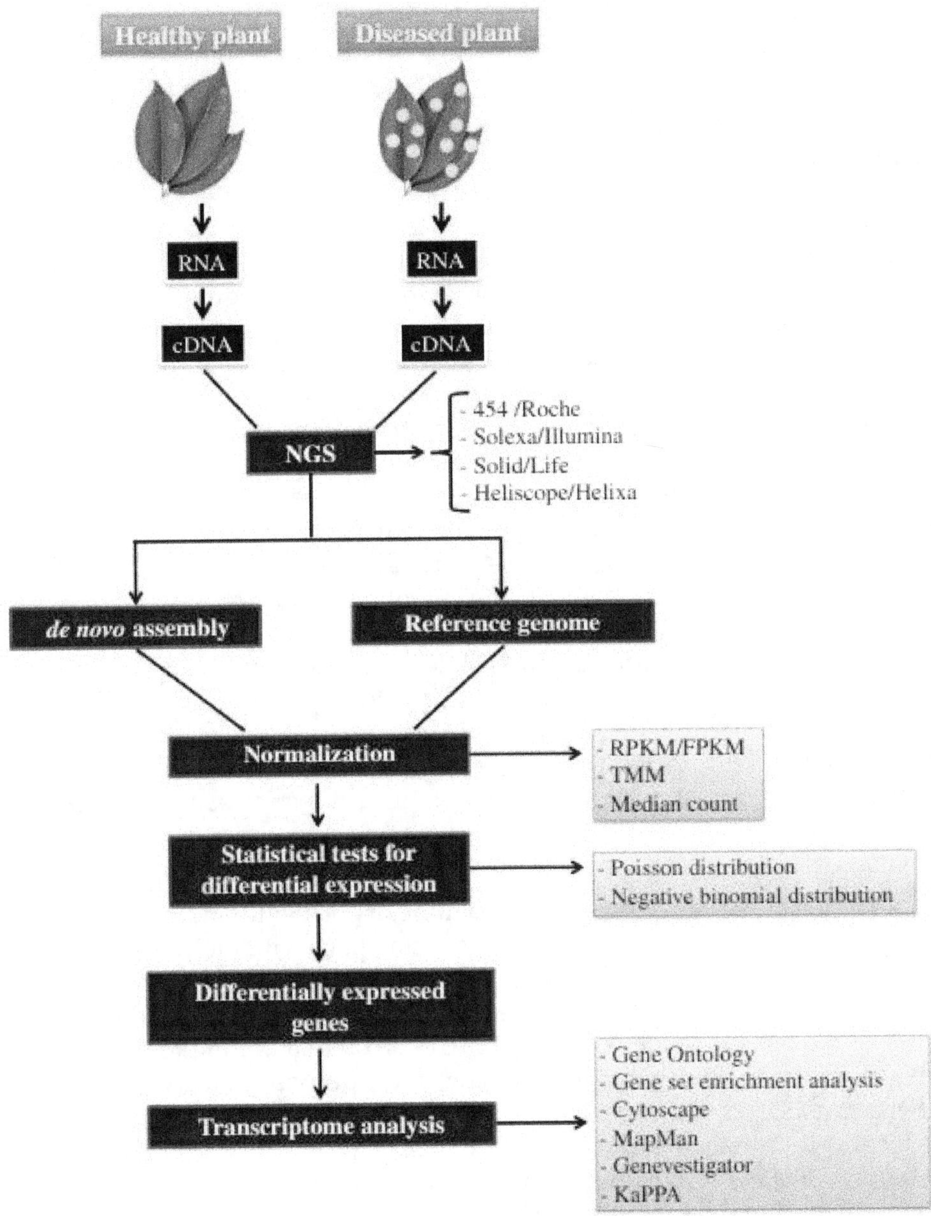

Figure 5.2: Schematic Representation of Gene Expression Analysis using RNA-seq.
RNA is isolated from two contrasting samples (healthy and diseased plant), and used for the synthesis of cDNAs. These are sequenced using one of the NGS methods, generating millions of reads for each sample. The reads are mapped against a reference genome or de novo assembly is performed. After this step, the data are normalized, and then statistical tests are performed to identify differentially expressed genes. Certain programs can infer the biological functions of these genes. (See color figure in color plate section). (Borem, 2014)

3. Serial Analysis of Gene Expression (SAGE)

SAGE was the first sequencing-based method to quantify the profusion of thousands of transcripts simultaneously (Morozova *et al.*, 2009). It is based on the principle that a short DNA sequence (tag) of 9–11 bp, derived from a defined location within one transcript, contains enough information to uniquely identify that transcript, if the position of the sequence within the transcript is known. By counting the tags, SAGE provides an estimate of the abundance of the transcripts (Vesculescu *et al.*, 1995). Briefly, the method involves the generation of a library of clones containing concatenated short sequence tags from a population of mRNA transcripts. These clones are sequenced using standard Sanger sequencing; tag sequences are matched to reference sequences in other databases to recognize the transcripts, and the tags are counted to estimate the comparative richness of the corresponding transcripts. The frequency of these transcripts in two or more SAGE libraries can be associated to distinguish dissimilarities in expression of gene in the respective samples (Madden *et al.*, 2000). SAGE offers some advantages compared with other methods, such as the EST sequencing and microarrays. In SAGE, instead of a clone each cDNA fragment representing only one transcript in a plasmid vector, multiple tags derived from different transcripts are exist in a unique plasmid vector. This strategy improves the throughput of data generated per sequencing run and reduces costs, compared with EST sequencing (Vega-Sánchez *et al.*, 2007). Unlike the microarray technology, which relies on previously identified sequences, SAGE has the ability to discover novel transcripts and to detect poorly expressed transcripts. By counting tags, SAGE obtains a direct measure of transcript abundance. The data obtained can be easily compared between multiple samples and across different experiments (Morozova *et al.*, 2009).

The major drawback of SAGE is the difficulty of identifying and annotating tags unambiguously, because short tags often match with multiple genes with similar coding sequences (Morozova *et al.*, 2009). In addition, reliable annotation depends on the existence of comprehensive EST or genomic databases to be used as reference (Madden *et al.*, 2000). To overcome some limitations of SAGE, modifications to the original methodology were proposed, such as LongSAGE (Saha *et al.*, 2002; Wei *et al.*, 2004) and SuperSAGE (Matsumura *et al.*, 2003), which result in tags with 21 and 26 bp, respectively, and DeepSAGE, which uses LongSAGE- derived ditags and replaces the Sanger sequencing with 454 pyrosequencing (Nielsen *et al.*, 2006).

In the past, SAGE has been applied extensively in medical research to profile the transcriptome of a range of human diseases, including cancer (Zhang *et al.*, 1997; Nacht *et al.*, 1999). However, since the first report of SAGE in plant research (Matsumura *et al.*, 1999), this method has not been widely used by plant biologists compared with other methods of studying large-scale transcriptomes, such as EST sequencing, microarrays, and more recently, RNA-seq.

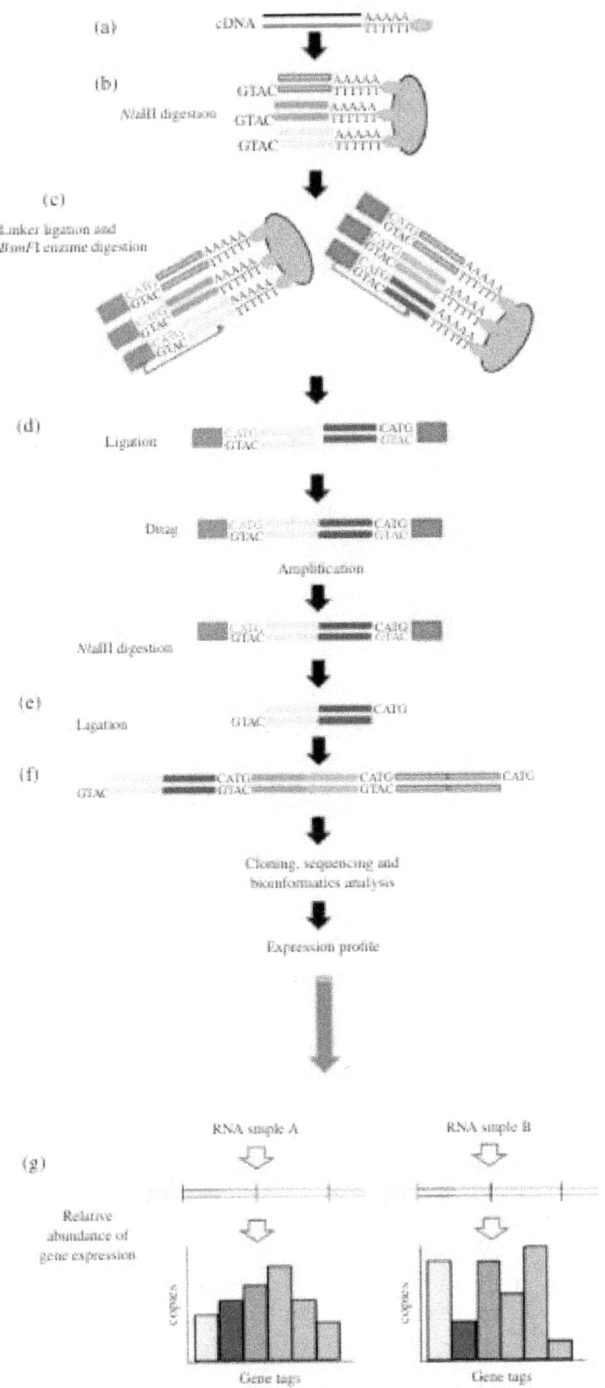

Figure 5.3: Schematic Representation of Serial Analysis of Gene Expression (SAGE) (Borem, 2014).

4. Construction of cDNA Libraries for EST Sequencing

ESTs are short sequence reads derived from the partial sequencing of complementary DNA (cDNA) sequences. The term "EST" was first proposed by Adams and co-workers in 1991. These researchers constructed a library of cloned cDNAs derived from transcripts of the human brain and generated ESTs by partial sequencing of 600 randomly selected clones, resulting in the discovery of new human genes (Adams *et al.*, 1991). This first report was the starting point for generating EST databases for myriad organisms. For decades, the information generated by ESTs was the main resource used to recognize new gene transcripts and assess gene expression levels in certain biological context.

Basically, the assembly of a cDNA library and generating ESTs involve: the isolation of total RNA and purification of mRNAs; synthesis of cDNA using a reverse transcriptase enzyme; cloning of cDNA fragments and random sequencing of selected clones using the Sanger method (Sanger *et al.*, 1977). Both the 32 and 52 ends of a cDNA clone could be sequenced, resulting in ESTs ranges from 100 to 700 bp in size (Nagaraj *et al.*, 2006).

After generating ESTs, the next step is EST sequence analysis, which is divided into pre-processing, clustering/assembly, and EST annotation. During pre-processing, vector adaptors and primer sequences are removed from the ESTs, and low quality and very short ESTs are also discarded from the dataset. Low complexity regions, such as repetitive elements and Simple Sequence Repeats (SSR) should also be detected and masked (replacing nucleotides of these regions with "N"). Poly(A)-tails should be trimmed to retain a few adenines (Nagaraj *et al.*, 2006). Pre-processing results in high-quality ESTs suitable for the next step: the clustering and assembly of ESTs.

Several ESTs can be produced from the same cDNA; therefore, ESTs can be divided into groups (clusters) according to their sequence similarity. Subsequently, each cluster is assembled to generate a consensus sequence, namely, a contig. More than one contig can be generated for each cluster. Sequences that cannot be grouped with other ESTs are called singletons, which may represent rare transcripts or contamination. The next step, the annotation of contigs, can be performed by sequence similarity searches comparing DNA or protein sequences deposited in public databases, such as the GenBank. BLAST algorithms from the NCBI (National Center for Biotechnology Information) are used for similarity searches of ESTs against nucleotide (BLASTN) or protein (BLASTX) sequence databases (Altschul *et al.*, 1990). Based on the sequence similarity of contigs with genes of model organisms, functional annotation can also be performed. Databases for functional annotation include, for example, GO (Gene Ontology) and KEGG (The Kyoto Encyclopedia of Genes and Genomes), which provide information about ontologies (biological processes, cellular components, and molecular functions) and metabolic pathways, respectively.

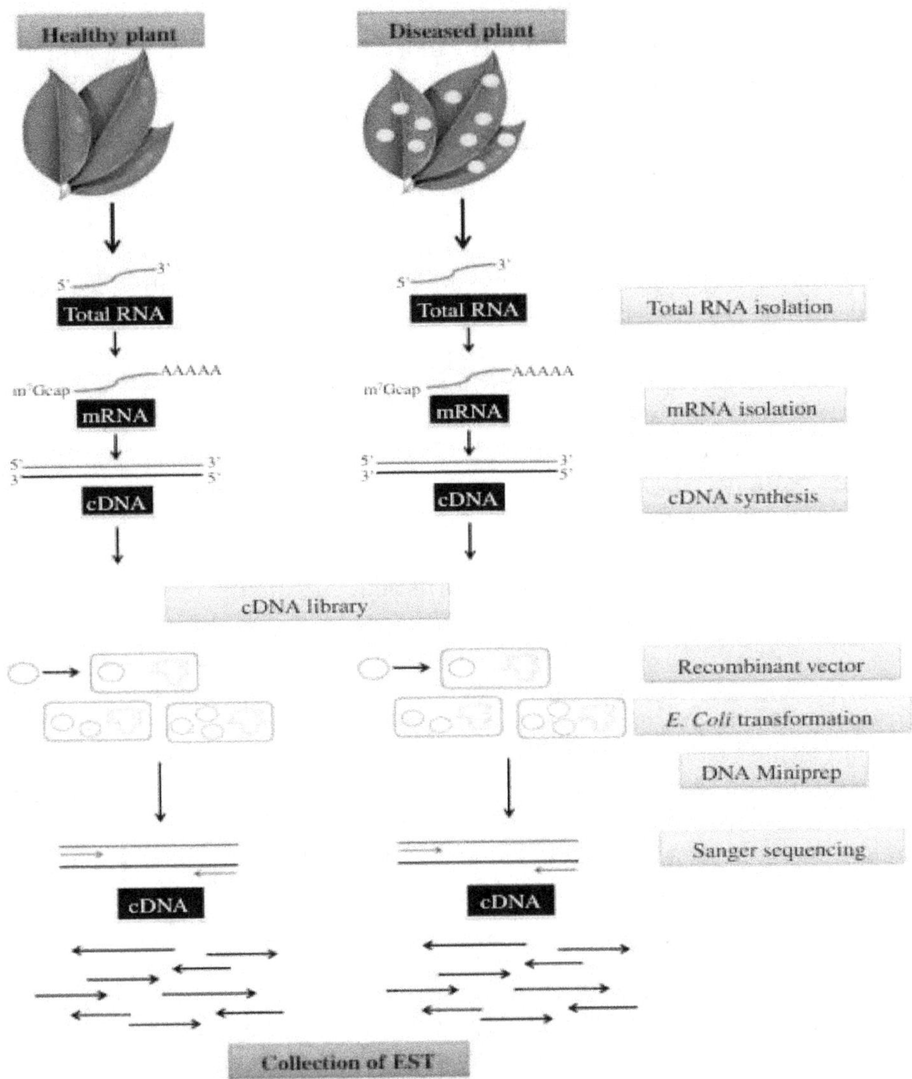

Figure 5.4: Overview of cDNA Construction and ESTs Generation.

The process starts with the isolation of mRNA from the total RNA individually isolated from samples of interest (in this example, leaf samples of diseased and healthy/control plants). mRNAs are reverse transcribed to produce complementary DNA (cDNAs) creating libraries of cDNAs can be cloned into an suitable vector. After cloning and E. coli transformation, specific clones from these libraries are randomly selected and subjected to a single sequencing reaction using universal primers; one or both ends of the cloned fragment (insert) are sequenced. The collection of short fragments (ESTs) generated is then processed using a no. of bioinformatics tools. In this example, it is also possible to use the relative abundance of ESTs representing the identical gene to compare the level of gene expression between two experimental conditions. (See color figure in color plate section). (Borem, 2014).

Collections of ESTs generated by the sequencing of cDNA libraries are stored in EST database repositories. The first database created for EST sequences was dbEST (Boguski *et al.*, 1993). Other public databases include EMBL (Stoesser *et al.*, 2003), which also archives all available ESTs, and UniGene (Boguski and Schuler, 1995; Schuler *et al.*, 1996), which contains clustered EST sequences retrieved from dbEST. Since early 2000s, EST sequencing has been widely used to survey the transcriptome of many plant species. This method is a quick and low- cost alternative to whole genome sequencing, providing a valuable resource for gene discovery, genome annotation, and comparative genomics, especially for plant species with large, complex genomes. High-throughput EST sequencing has also been performed to complement genome sequencing projects, to identify many polymorphisms, and for developing gene-based molecular markers (Barbazuk *et al.*, 2007). Despite the useful and valuable information provided by EST data, EST sequencing posses some limitations. ESTs are sequenced only once; therefore, they are often of low quality and may have a high error rate. These errors result from substitutions, deletions, and insertions of nucleotides when compared with the original mRNA sequence. EST libraries are also subject to contamination by sequences generated from genomic DNA, vector DNA, chimeric cDNAs (artifacts produced during ligation and reverse-transcription reaction), mitochondria or ribosomal DNA (Nagaraj *et al.*, 2006). Poorly expressed transcripts are often missed or under-represented within libraries, while highly expressed transcripts are over- represented. If an EST is not represented within an EST library, it does not mean that the corresponding gene is not expressed or absent from the genome. Moreover, EST sequencing by the Sanger method is laborious, time-consuming, and remains expensive for surveying large-scale transcriptomes. Thus, EST data are not suitable for estimating transcript abundance (Morozova *et al.*, 2009).

5. Suppression Subtractive Hybridization (SSH)

Suppression subtractive hybridization (SSH) is a method used for separating either DNA or cDNA molecules that distinguish two related samples. In particular, the SSH technique can be applied to study transcriptomics, as it is a comparative method that examines the relative richness of transcripts of a sample of interest (tester) in relation to a control sample (driver) (Luk'ianov *et al.*, 1994; Diatchenko *et al.*, 1996; Gurskaya *et al.*, 1996). SSH is based on the principle of PCR suppression, which allows the amplification of desired sequences andsimultaneously suppresses the amplification of undesirable ones. In addition, SSH combines normalization and subtraction during the hybridization step, which removes cDNAs that are common between the tester and driver samples and normalizes (equalizes) cDNAs with different concentrations. Normalization occurs because during hybridization, more abundant cDNA molecules can anneal faster than less abundant cDNA fragments that remain single-stranded (Lukyanov *et al.*, 2007). Finally, the method retains only differentially expressed or variable sequence transcripts that were present in the tester (Desai *et al.*, 2001).

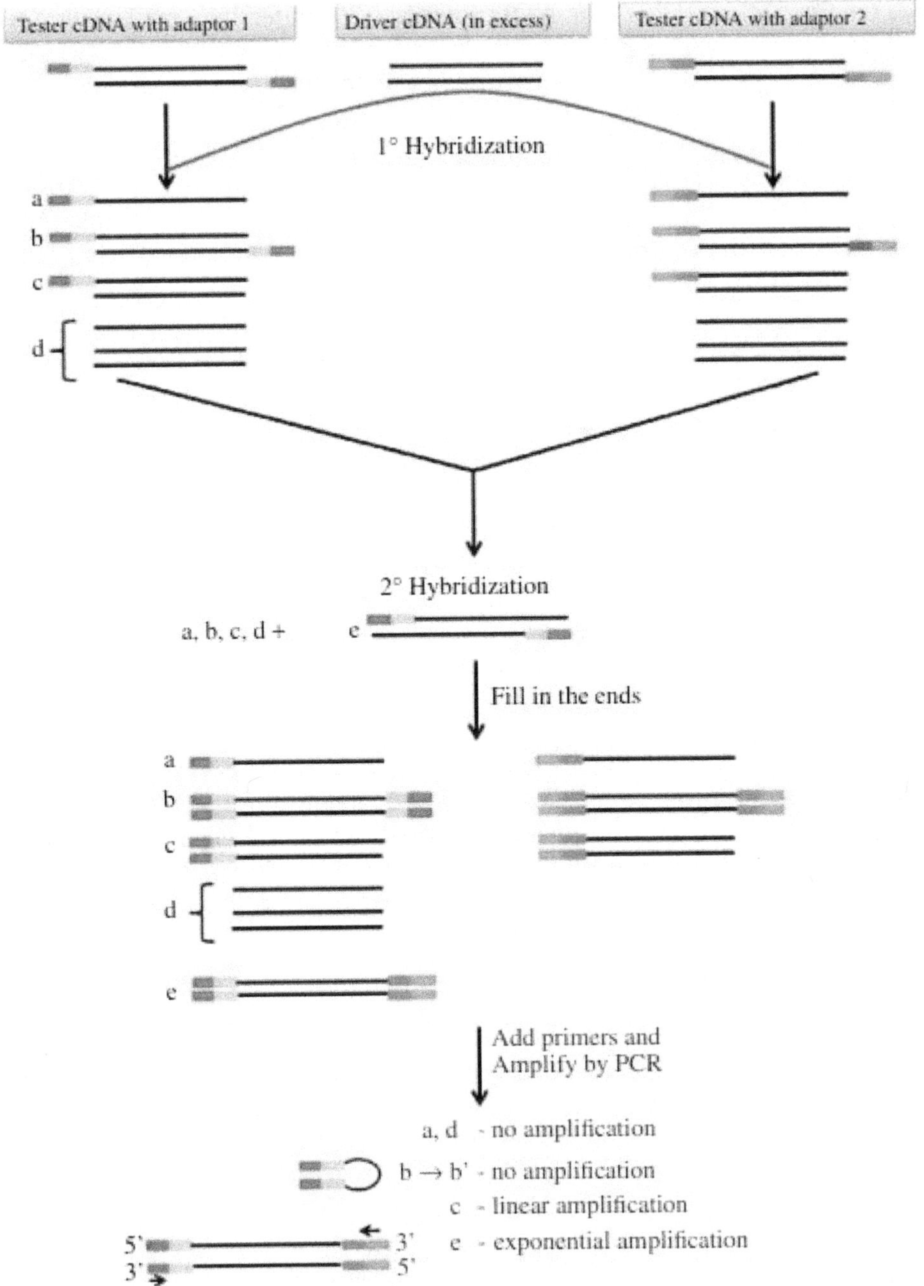

Figure 5.5: Scheme of the Suppression Subtractive Hybridization Method. (Borem, 2014).

Briefly, the SSH technique starts with RNA isolation from the tester and driver samples, followed by mRNA isolation and cDNA synthesis. Tester and driver cDNA samples are digested with a blunt cutting restriction enzyme, and the resulting tester cDNA fragments are then subdivided in two aliquots. Each aliquot is ligated with a different adapter (adapters 1 and 2), which contain self-complementary ends. After two rounds of hybridization, only target double-stranded fragments that are present in the tester and have different ligated adapters can be exponentially amplified. After denaturing and annealing, single-stranded fragments containing the same adapter at both ends form a stem–loop structure and their amplification is suppressed. After confirming that a subtracted sample contains an abundance of cDNAs originating from genes with differential expression, the next step involves cloning or subcloning these fragments to produce a subtracted cDNA library. Subsequently, selected clones are sequenced and the functional annotation of sequences can be performed using the same procedure described in the EST sequencing method.

SSH technique is a powerful approach for detecting differential gene expression of organisms with no previous information of their genome or transcriptome. The procedure has the benefit of eliminating physical parting of ss- and ds-cDNAs, requiring only one round of subtractive hybridization. The two hybridization steps lead to an efficient normalization of cDNA concentrations, achieving over 1000-fold enrichment for differentially expressed cDNAs, including rare transcripts. However, in practice, the level of enrichment depends on several factors, such as the concentration of tester relative to the driver sample, the no. of differentially expressed genes, the nature and complexity of the samples used, and the time of hybridization (Lukyanov *et al.*, 2007). Also, SSH techniques can generate many false positives, which are undesirable (background) clones representing non-differentially expressed genes. In some cases, the number of false positives can exceed the number of target clones in the subtracted library. To overcome this problem, a modified method, known as "mirror orientation selection" can be used to decrease the number of false positive clones (Rebrikov *et al.*, 2000).

6. Single Cell RNA Sequencing

Single-cell RNA sequencing (scRNA-seq) is a powerful molecular biology technique that allows researchers to analyze gene expression profiles at the single-cell level. Traditional bulk RNA sequencing measures gene expression levels from a mixture of cells, providing an average representation of gene expression across the entire cell population. However, this approach overlooks the inherent variability and heterogeneity present within a cell population.

ScRNA-seq overcomes this limitation by providing insights into the gene expression patterns of individual cells within a sample. This technique enables the identification of distinct cell types, subpopulations, and rare cell states that might be obscured in bulk sequencing data. It has revolutionized our understanding of cellular heterogeneity, developmental processes, disease mechanisms, and more.

The basic workflow of sc RNA-seq involves the following steps:

1. **Cell Isolation** Cells are isolated from the biological sample of interest. Depending on the experimental design, cells can be obtained from tissues, blood, or other sources.

2. **Single-Cell Isolation:** Individual cells are isolated and captured in separate reaction wells or microfluidic droplets. This step ensures that each cell's RNA content can be analyzed individually. **RNA Extraction:** RNA is extracted from each isolated cell. This RNA will be used to analyze gene expression.

3. **Library Preparation:** The extracted RNA is converted into cDNA (complementary DNA), and specific barcodes are added to each cDNA molecule to keep track of the cell of origin. This step is essential for distinguishing individual cells during data analysis.

4. **Sequencing:** The cDNA libraries are sequenced using high-throughput sequencing technologies. This generates a large amount of short sequence reads that represent the profile of gene expression for each individual cell.

5. **Data Analysis:** The sequencing data is then processed and analyzed to identify the genes that are expressed in each cell. Various computational methods are used to cluster cells into different subpopulations based on their gene expression patterns. Dimensionality reduction techniques, such as principal component analysis (PCA) and t-distributed stochastic neighbor embedding (t-SNE), are commonly used to visualize and explore the high-dimensional data.

6. **Cell Type Identification:** By comparing gene expression patterns with known marker genes or through more sophisticated computational methods, researchers can identify different cell types, states, or subpopulations within the sample.

Prospects of Transcriptomics

Transcriptomics to Study Abiotic Stress Tolerance in Plants

As the number of whole-genome transcriptomic investigations in plants grows, there is a tendency to overlook the genes associated with stress response, downstream signaling, and the synthesis of stress-responsive molecules. A wealth of data on transcriptomics in cereal crops, including rice, sorghum, wheat, maize and barley, has become accessible. This data has provided us with valuable insights into how different biological processes are synchronized in various plant tissues when they encounter stressful conditions. Studying the effects of drought stress on a plant's flowering or fruiting stages offers valuable information about the interplay within the reproductive system, the signaling of hormones, and the pathways of metabolism.

By comparing the transcriptomes of drought-tolerant and susceptible cultivars, potential candidate genes are identified and the mechanisms of adaptation under drought stress are illuminated. Previous studies have revealed the upregulation of 20 CIPK genes in rice, particularly in response to drought conditions. Recent RNA-seq investigations have further highlighted the enhanced expression of these CIPK genes under various abiotic stresses, including salinity and cold stress. RNA-seq analyses have provided evidence that salt-tolerant rice cultivars mount rapid responses to salinity, involving early induction of H2O2 and prompt signal transduction compared to their sensitive counterparts. Salinity-tolerant strains employ an adaptive strategy by containing sodium within roots and mature leaves, while simultaneously activating genes associated with photosynthesis in emerging leaves.

In a comprehensive transcriptomics study, two inbred lines with contrasting cold tolerance levels unveiled 948 differentially expressed genes (DEGs) among 19,794 total genes. These DEGs predominantly relate to DNA binding, ATP binding, and protein kinase functions.

In a separate RNA-seq analysis, involving drought-resistant and drought-susceptible sorghum seedlings subjected to PEG-induced drought stress, 180 DEGs emerged, including 70 novel genes associated with transcription factors and stress signal transduction.

Utilizing Transcriptomics for Enhancing Crop Resilience to Biotic Stress

Crop yield faces numerous challenges from biotic stress factors, including bacteria, viruses, fungi, insect pests, and weeds. "Many plant breeding efforts focus on developing genotypes resistant to plant pathogens and insect pests to reduce crop loss due to biotic stress. Plants have evolved a range of mechanisms, including salicylic acid and reactive oxygen species production, to combat biotic stressors. Understanding the molecular changes in plants when dealing with pathogen invasion is essential for enhancing disease-resistant crop varieties.

Numerous transcriptomic investigations have been carried out in cereals to decode disease resistance mechanisms and identify resistance (R) genes. Examining the complete genome transcriptome of four wheat varieties (Wuhan 1, Nyubai, HC374, and Shaw) following Fusarium graminearum head inoculation revealed the increased expression of leucine-rich repeat receptor kinases (LRR-RKs). These kinases play a pivotal role in disease resistance and exhibited varying expression levels at different time intervals in both resistant and susceptible cultivars. These distinct expression patterns in different genotypes demonstrated unique genotype-specific defense responses. High-quality transcriptomic investigations through RNA-seq provide valuable insights into the genomic dynamics within host-pathogen systems.

The rice Xa23 gene is renowned for its broad-spectrum resistance against various biotypes of Xanthomonas oryzae pv. oryzae (Xoo). Transcriptome profiling of Near Isogenic Lines (NILs), one carrying Xa23 (CBB23) and the other lacking it (JG30), before and after Xoo inoculation, provides insights into the downstream genes and pathways involved in Xa23-mediated resistance. A total of 1645 differentially expressed genes (DEGs) were identified, with a substantial proportion linked to phenylpropanoid biosynthesis, followed by flavonoid biosynthesis and phytohormone signaling.

REFERENCES

Alexa, M.D., Kelley, J M., Gocayne, J D., *et al.*, 1991. Complementary DNA sequencing: expressed sequence tags and human genome project. Science, 252 (5013): 1651–1656.

Alexa, A., Rahnenführer, J., Lengauer, T. 2006. Improved scoring of functional groups from gene expression data by decorrelating GO graph structure. Bioinformatics, 22 (13): 1600–1607.

Altschul, S.F., Gish, W., Miller, W., *et al.*, 1990. Basic local alignment search tool. Journal of Molecular Biology, 215 (3): 403–410.

Anjum, N. A., Gill, S. S., Ahmad, I., Tuteja, N., Soni, P., Pareek, A., and Pereira, E. (2012). Understanding StressResponsive Mechanisms in Plants: An Overview of Transcriptomics and Proteomics Approaches. *Improving crop resistance to abiotic stress*, 337-355.

Barbazuk W.B., Emrich, S.J., Chen, H.D., *et al.*, 2007. SNP discovery via 454 transcriptome sequencing. The Plant Journal, 51 (5): 910–918.

Batista, R., Saibo, N., Lourenço, T., and Oliveira, M. M. (2008). Microarray analyses reveal that plant mutagenesis may induce more transcriptomic changes than transgene insertion. *Proceedings of the National Academy of Sciences*, 105(9), 3640-3645.

Boguski, M.S., Schuler, G.D. 1995. ESTablishing a human transcript map. Nature Genetics, 10 (4): 369–371.

Borem, Aluizio, Fritsche-Neto, Roberto (2014). *Omics in Plant Breeding (Borem/ Omics in Plant Breeding) Transcriptomics 10.1002/9781118820971 (), 33-57.* doi: 10.1002/978111882097.ch3

Coram, T. E., Brown-Guedira, G., and Chen, X. M. (2009). Using transcriptomics to understand the wheat genome. *CABI Reviews*, (2008), 1-9.

Desai, S., Hili, J., Trelogan, S., *et al.*, 2001. Identification of differentially expressed genes by suppression subtractive hybridization. In: Stephen, H. and Levesey, R. (eds). Functional Genomics: A Practical Approach. Oxford: Oxford, University Press, pp. 45–80.

Diatchenko, L., Lau, Y.F., Campbell, A.P., *et al.,* 1996. Suppression subtractive hybridization: a method for generating differentially regulated or tissue-specific cDNA probes and libraries. Proceedings of the National Academy of Sciences U.S.A., 93 (12): 6025–6030.

Grabherr, M.G., Haas, B.J., Yassour, M., *et al.,* 2011. Full-length transcriptome assembly from RNA-Seq data without a reference genome. Nature Biotechnology, 29 (7): 644–652.

Gurskaya, N.G., Diatchenko, L., Chenchik, A., *et al.,* 1996. Equalizing cDNA subtraction based on selective suppression of polymerase chain reaction: cloning of Jurkat cell transcripts induced by phytohemaglutinin and phorbol 12-myristate 13-acetate. Annual Biochemistry, 240: 90–97.

Harris, M.A., Clark, J., Ireland, A., *et al.,* 2004. The Gene Ontology (GO) database and informatics resource. Nucleic Acids Research, 32 (Database issue), D258.

Kaur, B., Sandhu, K. S., Kamal, R., Kaur, K., Singh, J., Röder, M. S., and Muqaddasi, Q. H. (2021). Omics for the improvement of abiotic, biotic, and agronomic traits in major cereal crops: Applications, challenges, and prospects. *Plants, 10*(10), 1989.

Luk'ianov, S.A., Gurskaia, N.G., Luk'ianov, K.A., *et al.,* 1994. Highly-effective subtractive hybridization of cDNA. Bioorganicheskaya khimiya, 20: 701–704.

Lukyanov, S.A., Rebrikov, D., Buzdin, A.A. 2007. Suppression subtractive hybridization. In: Buzdin, A.A. and Lukyanov S.A. (eds). Nucleic Acids Hybridization Modern Applications. Dordrecht, The Netherlands: Springer, pp. 53–84.

Madden, S. L., Wang, C.J., Landes, G. 2000. Serial analysis of gene expression: from gene discovery to target identification. Drug Discovery Today, 5 (9): 415–425.

Matsumura, H., Reich, S., Ito, A. *et al.* 2003. Gene expression analysis of plant host–pathogen interactions by SuperSAGE, Proceedings of the National Academy of Sciences U.S.A., 100: 15718–15723

Morozova, O., Hirst, M., Marra, M.A. 2009. Applications of new sequencing technologies for transcriptome analysis. Annual Review of Genomics and Human Genetics, 10: 135–151.

Nacht, M., Ferguson, A.T., Zhang, W., *et al.,* 1999. Combining serial analysis of gene expression and array technologies to identify genes differentially expressed in breast cancer. Cancer Research, 59: 5464–5470.

Nagalakshmi, U., Wang, Z., Waern, K., *et al.,* 2008. The transcriptional landscape of the yeast genome defined by RNA sequencing. Science, 320: 1344–1349.

Nagaraj, S.H., Gasser, R.B., Ranganathan S. 2006. A hitchhiker's guide to expressed sequence tag (EST) analysis. Briefings in Bioinformatics, 8 (1): 6–21.

Nielsen, K.L., Hogh, A.L., Emmersen, J. 2006. DeepSAGE–digital transcriptomics with high sensitivity, simple experimental protocol and multiplexing of samples. Nucleic Acids Research, 34: e133.

Oshlack, A., Robinson, M.D., Young M.D. 2010. From RNA-seq reads to differential expression results. Genome Biology, 11: 220.

Rebrikov, D.V., Britanova, O.V., Gurskaya, N.G., *et al.*, 2000. Mirror orientation selection (MOS): a method for eliminating false positive clones from libraries generated by suppression subtractive hybridization. Nucleic Acids Research, 28: e90.

Saha, S., Sparks, A.B., Rago, C., *et al.*, 2002. Using the transcriptome to annotate the genome. Nature Biotechnology, 20: 508–512.

Sanger, F., Nicklen, S., Coulson, A.R. 1977. DNA sequencing with chain terminating inhibitors. Proceedings of the National Academy of Sciences U.S.A., 74: 5463–5467

Schuler, G.D., Boguski, M.S., Stewart, E.A., *et al.*, 1996. A gene map of the human genome. Science, 274: 540–546.

Shannon, P., Markiel, A., Ozier, O., *et al.*, 2003. Cytoscape: a software environment for integrated models of biomolecular interaction networks. Genome Research, 13 (11): 2498–2504.

Shendure, J., Ji, H. 2008. Next-generation DNA sequencing. Nature Biotechnology, 26 (10): 1135–1145

Soneson, C., Delorenzi, M. 2013. A comparison of methods for differential expression analysis of RNA-seq data. BMC Bioinformatics, 14 (1): 91.

Stoesser, G., Baker, W., van den Broek, A., *et al.*, 2003. The EMBL nucleotide sequence database: major new developments. Nucleic Acids Research, 31: 17–22.

Thimm, O., Bläsing, O., Gibon, Y., *et al.*, 2004. Mapman: a user-driven tool to display genomics data sets onto diagrams of metabolic pathways and other biological processes. The Plant Journal, 37 (6): 914–939.

Tokimatsu, T., Sakurai, N., Suzuki, H., *et al.*, 2005. KaPPA-View. A web-based analysis tool for integration of transcript and metabolite data on plant metabolic pathway maps. Plant Physiology, 138 (3): 1289–1300.

Trapnell, C., Williams, B.A., Pertea, G., *et al.*, 2010. Transcript assembly and quantification by RNA-Seq reveals unannotated transcripts and isoform switching during cell differentiation. Nature Biotechnology, 28 (5): 511–515.

Vega-Sánchez, M.E., Malali, G., Guo-Liang, W. 2007. Tag-based approaches for deep transcriptome analysis in plants. Plant Science, 173 (4): 371–380.

Velculescu, V.E., Zhang, L., Vogelstein, B., Kinzler, K.W. 1995. Serial analysis of gene expression. Science-AAAS-Weekly Paper Edition, 270 (5235): 484–486.

Wang, Z., Gerstein, M., Snyder, M. 2009. RNA-Seq: a revolutionary tool for transcriptomics. Nature Reviews Genetics, 10 (1): 57–63.

Ward, J. A., Ponnala, L., and Weber, C. A. (2012). Strategies for transcriptome analysis in nonmodel plants. *American journal of botany*, 99(2), 267-276.

Wei, C.L., Ng, P., Chiu, K.P., *et al.*, 2004. 52 Long serial analysis of gene expression (LongSAGE) and 32 LongSAGE for transcriptome characterization and genome annotation. Proceedings of the National Academy of Sciences U.S.A., 101: 11701– 11706.

Zhang, L., Zhou, W., Velculescu, V.E., *et al.*, 1997. Gene expression profiles in normal and cancer cells. Science, 276: 1268–1272.

Zhu, T. (2003). Global analysis of gene expression using GeneChip microarrays. Current opinion in plant biology, 6(5), 418-425.

Zimmermann, P., Hirsch-Hoffmann, M., Hennig, L., Gruissem, W. 2004. GENEVESTIGATOR. Arabidopsis microarray database and analysis toolbox. Plant Physiology, 136 (1): 2621– 2632.

Chapter 6

Recent Trends in Proteomics and Metabolomics

Jwala Pranati[1], Maulik R. Prajapati[2], Suma K.[3], Swatirekha N.[4] and Arun Zacharia[3]*

[1]*Professor Jayashankar Telangana State Agricultural University, Rajendranagar, Hyderabad – 500 020*
[2]*Navsari Agricultural University, Navsari – 396 450, Gujarat*
[3]*University of Agricultural Sciences, Bengaluru – 560 065, Karnataka*
[4]*Acharya N.G Ranga Agricultural University, Bapatla – 522 034, Andhra Pradesh*
**e-mail: jwalapranati1601@gmail.com*

ABSTRACT

Present day world has reached a crossroads to strategically and immediately address the need to feed the 9 billion people by 2050, along with tackling climate change. Development in biology and plant-omics take a major place in sharing the progress to address the issues. Proteomics and metabolomics techniques are the recent resorts to connect the chemical phenotype to genotype for by understanding the gene protein ability, its characterization, post-translational modification, metabolites involved in the signaling pathway and utilizing them as biomarkers for easy selection. This chapter thus focusses on briefing the proteomic and metabolomic advances in plants and how they are being leveraged for benefit.

INTRODUCTION

Currently global warming and population explosion have become the most concerning phrases around the world impacting the entire biosystem directly or indirectly. Efforts are underway in this direction to modify the agricultural production and productivity to answer these situations. With the advent of molecular breeding tools such as markers, sequencing platforms, bioinformatic databases it has eased the possibilities of targeting specific genes in the plant, identifying and characterizing them to breed for the future-ready crops. Omics technologies form one of the growing fields in adding molecular diagnosis to the conventional breeding platforms under which proteomics and metabolomic techniques aid to dig the gene protein and metabolites involved in the specific pathway thus connecting the chemical phenotype to genotype.

Proteome refers to 'the entire set of protein complement expressed by a genome in a cell or tissue type' and the systematic study of it is termed as proteomics (Rothemund *et al.*, 2003). Therefore, proteomics is the study of proteins, their function, structure, interactions, composition and their cellular activities. Marc Wilkins coined the term proteome in 1995 (Agarwal *et al.*, 2013). After the completion of the Human Genome Project, there was a gradual and steady shift from genomics to proteomics. With the advent of Next Generation Sequencing technologies, sequencing of the entire genome of organisms has become relatively simple and efficient which fastened the protein level understanding in plants.

A trait in a plant is expressed based on the genetic constitution and an environmental influence over the gene, eventually signaling a cascade of metabolomic reactions in the cell. These reactions lead to the production of low molecular weight molecules together termed as metabolome and this study is termed as "Metabolomics" by Steven Oliver. Comprehensive characterization of the primary and secondary metabolites or metabolome using advanced analytical chemistry technologies is called metabolomics (Wishart *et al.*, 2007). Though its yet an emerging science with major efforts still placed around the genomics and proteomic studies in major crops, the roots for the metabolite analysis dates back to 1817 to isolate chlorophyll by Bienaime Caventou (Delepine *et al.*, 1951). Since then, progress has been seen to analyze several metabolites including several analytical techniques such as gas chromatography (GC), liquid chromatography (LC), and mass spectrometry (MS) technologies (Holland *et al.*, 1986). The first successful attempts came up in 1991 to understand the herbicidal impact on the metabolome using GC which was extended to understand the macromolecules through high-throughput mass spectrometry with GC (Roessner *et al.*, 2000). Here, we summarise the brief working and advances in proteomic and metabolomic techniques in plants which could aid in breeding future-ready plants.

Advances in Proteomics

Extraction of Proteins, Purification and Separation Techniques

Isolation of proteins is the preliminary step in proteomics. The success of proteome analysis depends on the development of extraction procedures. Two basic methods are often used for diverse plant tissues - phenol extraction coupled with ammonium acetate precipitation and trichloroacetic acid precipitation (Isaacson *et al.*, 2006). TCA protocol is the most common extraction method used for complex plant tissues while solubilization of proteins in phenol is successfully used for recalcitrant species. Once the protein is extracted, their separation and purification are conducted through:

1. Gel based 2D-differential gel electrophoresis including SDS-PAGE
2. Gel free chromatography-based techniques such as high-performance liquid chromatography (HPLC)
3. Affinity purification based on interaction between affinity ligand of chromatographic matrix and protein sample utilized mainly to study the post translational modifications
4. Immunoprecipitation or pull-down approach to use generic or specific antibodies as bait to separate the proteins
5. Laser capture microdissection (LCM) to analyze gene protein expression and abundance from a cell culture of a specific cell or tissue of a plant body.

Of these methodologies, chromatography techniques have been dwelled to the maximum employing one dimensional or two-dimensional LC coupled with mass spectrometry for protein, PTM identification and characterization.

Quantitative or Descriptive Proteomics

Identification of different proteins in organisms necessitated the quantification of the identified proteins. Quantitative proteomics is the branch of proteomics which emphasizes the accurate abundance quantification of proteins and relative changes of protein expression in response to various changes (Ong and Mann, 2005). Quantification of protein samples based on their expression is done using reagents tandem mass tag (TMT) and Isobaric tags for relative and absolute quantification (iTRAQ). These tags label the free peptide groups obtained by fragmentation of the whole protein through mass spectrometry (MS) (Thompson *et al.*, 2003). Adding efficiency to the technique, the labels are re-fragmented to separate the isoforms (MS2, MS3). The fragments are computed through reverse phase high performance liquid chromatography (RPLC) for quantification of proteome.

Stable isotope labelling by amino acids in cell culture (SILAC) is another novel technique used in protein abundance profiling through metabolic labelling of proteome in a cell culture. Currently, its application is to understand the protein-protein interactions and relative protein levels in differentially labelled cultured (Lewandowska *et al.*, 2013).

Post Translational Modifications or Phosphoproteomics

Post translational modification (PTM) is an important phase of the proteome study which regulated numerous processes at cellular level. The major modifications include phosphorylation, ubiquitination, acylation and glycosylation (Ramazi and Zahiri, 2021). PTM create a change in mass of the proteome which can be analyzed through mass spectrometry (MS2, MS3). Thus, identification of proteoforms (different modifications through on the single protein) forms one of the major applications of proteomics. Phosphoproteomics studies occupy major focus in the PTM studies involving study of isoforms through phosphorylation of the whole protein. iTRAQ/TMT labelling followed by LC-MS analysis is extensively used for these studies.

Structural Analysis

Analysis of three-dimensional structure of proteins form the prerequisite for knowing the function and various interactions of proteins. Sequence analysis of the characterized proteins is important to further know its structure and functions. Edman sequencing is one of the techniques to decipher the sequence of amino acids in proteins. It is greatly used in biopharmaceuticals and in development of therapeutic proteins. X-ray crystallography is one of the important techniques for determining the 3-D structure of proteins, which uses X-ray to decipher the size of repeating units that forms the crystal. NMR spectroscopy is another tool to examine the molecular structure, folding pattern and behavioural changes of proteins. Structural analysis helps in knowing of mechanism of enzyme action, designing of drugs and site directed mutagenesis.

Analysis of Data Using Bioinformatics Tools

Bioinformatic tools are the central component to derive accurate inference from the information on protein sequence and structural analysis. They use novel algorithms to manage and utilize hegemony of heterogeneous protein data. Protein sequence databases used for proteome analysis are GenBank, RefSeq, nr, UniProt, UniRef, Ensembl, PlantGDB used for sequence retrieval. Yet to process the large datasets from mass spectrometry dedicated search tools such as Andromeda, Open MS, PEAKS, SILVER, RIPPER, PANDA Proteome discoverer, Protein Pilot, MassHunter, MASCOT, OMSSA, PhenyX, Protein Prospector, etc are utilized. These databases provide common platform for obtaining and processing the raw information and providing output on the protein identification, post translational modification, protein-protein interactions. Further the data obtained is submitted to the gene ontogeny databases such as KEGG, Ingenuity, Blast2GO, STRING, BIOGRID, SKYLINE to obtain the function of proteins.

Application of Proteomics

With the advent of sequencing technologies and genomic tools, multiple research paths are being explored to connect gene to the particular specific protein in the cell. The major focus relied currently on uncovering the mechanisms of stress tolerance in organisms, while there are few success stories in further exploring the protein-protein interactions and the single cell proteomics. A glimpse of work under these aspects is provided in Table 6.1.

Though, there is significant progress in the proteome understanding, there are difficulties considering the clear identification and separation of isobaric proteins which requires advanced chromatography techniques. Furthermore, improvement is the databases and software's utilized for data mining and interpretation would help minimize the false discovery rate (FDR). This technique provides a bridge for phenotype to protein unravelling, yet integrating the proteomics with other omics technologies would add to the benefit.

Advances in Metabolomics

All the molecular biology studies combined with omics technologies including genomics, transcriptomics, proteomics, and other molecular phenotyping techniques guide the metabolomics research (Scossa *et al.*, 2021). Mapping specific metabolites provides us insights on the signaling pathways from the gene to the end product level within, could be explored to tap the genetic gain for complex traits. Targeted and untargeted metabolomics are two of metabolomics analysis.

Sample Preparation, Separation and Identifying the Metabolites

Sample preparation is the most sensitive step in the metabolome analysis. Robust, reproducible method of extraction is to be chosen to extract the primary and secondary metabolites with highest turnover. Harvesting the sample and quenching with liquid nitrogen is essential for homogenization of the sample. One of the conventional techniques practiced for sample extraction is Soxhlet extraction method. Here, the sample is heated continuously, within a concentrated solvent used for extraction. Solid phase microextraction combines with mass spectrometry, Laser microdissection (LMD), Microwave-assisted extraction (MAE), supercritical fluid extraction (for volatile metabolites), ultrasound-assisted extraction (UAE), enzyme-assisted extraction, solid phase microextraction (SPME), and the Swiss rolling technique are few other techniques utilized for extraction of metabolome based on the analytic components.

Once the sample is prepared, various methods can be employed to separate and characterize groups of metabolites that are chemically distinct. Nuclear magnetic resonance (NMR) and Mass Spectrometry (MS) analysis are commonly used analytical platforms for data acquisition in metabolome analysis, and these platforms are frequently combined with separation techniques like gas chromatography (GC), high or ultraperformance liquid chromatography (HPLC

Table 6.1: Various Application in Proteomic Research

Application	Trait	Proteins	Role	Crop	References
Identification and quantification of the proteins responsible	Salt Stress Tolerance	LEA proteins	Seed and hypocotyl development	Soybean	Aghaei *et al.*, 2008
	Drought tolerance	Acyl-Co-A- binding protein	Upregulate plasma membrane proteins	Maize	Zeng *et al.*, 2019
	Nitrogen efficiency	Peroxidase	Root Adaptations	Rapeseed root	Qin *et al.*, 2019
	Cold stress	Kinases, ligases, transcriptional factors	Cytoskeleton reorganization, calcium signaling, MAPK cascades	Arabidopsis	Tan *et al.*, 2021
Understanding the protein-protein interactions and the signaling pathway through AP-MS	Seed germination	Germination 1(DOG1)	Phosphatases -AHG1, AHG3, RDO5 and PDF1 interact with DOG1 to control germination	Arabidopsis	Nee *et al.*, 2017
Single cell proteomics through LCM	Aluminium tolerance	313 differential abundance proteins	Antioxidant proteins increase in the outer layer cells of root	Tomato root	Potts *et al.*, 2022

or UPLC), or capillary electrophoresis (CE) analysis (Braga *et al.*, 2018). Each detection or separation technique is chosen in accordance with chemical and physical characteristics of each sample and the type of analysis to be carried out. Each one has a distinct resolution, sensitivity, and technological constraints in terms of identifying metabolites (targeted or untargeted).

Untargeted Metabolomics

This avenue of metabolomics takes up identification of global metabolome based on the chromatogram peaks in an unbiased analysis. It requires high resolution mass spectrometers for identification of isobaric co-eluting metabolites. Thus, models emerging currently utilize more than one chromatography techniques such as two-dimensional(2D) liquid chromatography (LC), 2D mass spectrometry (MS), 2D LC-MS.

Fluxomics is an emerging untargeted metabolomics approach which assesses the changes in metabolomic signaling pathways in the global metabolome using ^{13}C or ^{15}N labelling and analyzed through metabolic flux analysis (MFA).

Targeted Metabolomics

These subset choses the analysis of a particular class of metabolites using selective techniques. The techniques provide high resolution, rapid profiling of metabolites. GC-MS is the preferred technique for analysis of metabolites for large-scale analysis and identification of primary and secondary metabolites in the metabolome. Recently, LC-MS is emerging to be the next-in -line technique for targeted metabolomics.

Data Analysis and Annotation

Metabolomic analysis draws enormous data from analysis and the real dispute lies in computing the selective metabolites. The initial steps involve processing and statistical processing of data (Sun *et al.*, 2012) thus requiring sophisticated statistical tools. Metabolic marker probing is an emerging concept to connect the chemical phenotype with its respective metabolite biomarker. A pairwise Pearson's correlation, canonical correlation techniques can be computed to correlate the phenotype to the biomarker (Fernandez *et al.*, 2016).

Numerous statistical designs have been made available for metabolomic analysis to test the group wise differences either in a univariate, *i.e.*, parameter-by-parameter fashion using univariate techniques (t-test, analysis of variance (ANOVA), and Mann–Whitney U-test). Univariate analysis is generally performed for biomarker discovery at initial levels of systems biology, which studies one variable at a specific time.

Multiple data fusion strategies provide opportunity to explore the genotypic and phenotypic differences of plants (Fiehn *et al.*, 2018). There are numerous multivariate statistical tools including ANOVA, analysis of variance-simultaneous component analysis (A-SCA), principal component analysis (PCA), partial least squares-discriminant analysis (PLS-DA) and heat map analysis. The statistical tool applied depends on the experimental design and procedure (Ren *et al.*, 2018). Principal Component Analysis is one of the extensively used multivariate statistical approach for the multi-dimensional reduction approach. It is known for its work in dimensionality reduction and presenting in multiple principal components (Xu *et al.*, 2012).

While on the other end, the PLS method offers the benefit of handling erratic and strongly correlated data sets. Metabolic data analysis also commonly employs PLS extensions including orthogonal PLS (OPLS), sparse PLS (sPLS), and PLS-DA. The selection of metabolic markers can benefit from the important information provided by the OPLS and PLS approaches (Chun *et al.*, 2010, and Trygg *et al.*, 2002). MATLAB and SIMCA-P are two commercial statistical tools that offer a variety of methods.

Bioinformatics Tools and Databases

Numerous web-based applications have been created for metabolomics data mining, data assessment, data processing, and data interpretation as the field has expanded over the years. The online bioinformatics program XCMS helps the user with statistical analysis and data processing while allowing the user to upload raw data directly. The XCMS stream, recently established for planned data transfer in LC-MS experiments, has increased the effectiveness of an online system and cut down on data processing time (Tautenhahn *et al.*, 2012). By utilizing MS techniques and the METLIN database, it can also be utilized to identify mutative chemicals (Montenegro-Burke *et al.*, 2017). Few other databases used include METLIN, MetaGeneAlyse, MeltDB, iMet-Q, MS-Dial and MetAlign. MZedDB and KEGG have been widely used to research the metabolome with a species-nonspecific or species-specific origin (Draper *et al.*, 2009).

To study untargeted metabolites using LC-MS methods, a new tool called Galaxy-M has recently been created (Davidson *et al.*, 2016). Two omics-based web tools that have been used to perform univariate and multivariate statistical analysis, interpret gene expression data, and visualize metabolomics data are Meta box and Reich *et al.* (2006) Meta box, which is another online server with many applications in the elucidation of omics data.

Applications of Metabolomics

With the advent of new technologies, use of metabolomics has increased. It helps in dissecting many biological pathways and understanding them better. Role of primary and secondary metabolites in the growth and development of plants can be understood better. Major focus on metabolome analysis was vested in understanding the stress tolerance in plants and few of the examples are quoted in Table 6.2.

Table 6.2: Various Application of Metabolomic Research

Trait	Plant	Sample	Annotated Metabolites	Platform	References
Heat Stress	Soybean	Seed	Amino acids or amino acid derivatives, fatty acids, sugars, peptides, tocopherols, flavonoids, phenylpropanoids and ascorbate precursors	LC-LTQ/GC-MS	Chebrolu *et al.* (2016)
	Tomato	Pollen	Alkaloids, polyamines, flavonoids.	RPLC-qT OF	Paupiere *et al.* (2017)
Drought stress	Rice	Leaf	Sugar and sugar derivatives	GC-MS	Ma *et al.* (2016)
	Wheat	Root	Lipids, sugars, oxidative stress compounds and phytohormones	RPLC-qT OF	Yadav *et al.* (2019)
	Chickpea	Shoot	Phytol and GABA	GC-qTOF/ LC-qTOF	Herzog *et al* (2018)
	Barley	Root, Leaf	Isoflavones	1H-NMR	Coutinho *et al* (2018)
Flooding stress	Rice	Leaf	Proline, lysine, alanine and GABA	GC/MS	Che-Othman *et al.* (2019)
	Wheat	Roots and Shoots	Malic acid, proline, fructose, mannose, glycine, Glutamic acid	HPLC	Borrelli *et al.* (2018)
Salt Stress	Maize	Leaf	Glycine, alanine and GABA	GC/MS	Barding *et al.* (2012)
	Barley	Leaf	6-phosphogluconate, phenylalanine and lactate	GC/MS and NMR	Locke *et al.* (2018)
	Tomato	Roots and leaves	Hydroxycinnamic acids and glucosinolates	NMR	Jahangir *et al.* (2008)
Water logging	Rice	Roots and leaves	Fatty acids	capHPLC-ESI (−)-QTOF-MS	Ibarra *et al.* (2019)
	Soybean	Leaf	Fucose, ribulose, lyxose, galactinol and erythrito	GC-TOF-MS	Heyneke *et al.* (2017)
Metal stress	Canola	Flag leaf	Methylisoorientin-2"- O-rhamnoside, iso-orientin and iso-vitexin	UPLC-QTOF	Zhang *et al.* (2017)
	Sunflower	Roots and leaves	Sulfur metabolites, organic acids and amino acids	UPLC	Ghosson *et al.* (2018)

Biotic Stress	Plant	Sample	Annotated Metabolites	Platform	References
Zymoseptoria tritici	Wheat	Leaf	Flavonoids, hydroxycinnamic acid amides and cinnamyl alcohols	FT-ICR- MS	Boiteau *et al.* (2018)
Fusarium graminearum	Wheat	Rachis and spikelet	Fatty acids, terpenoid, phenolic glycosides, flavonoid and phenyl propanoids	LC-LTQ-Orbitrap	Gunnaiah *et al.* (2012)
Orseolia oryzae	Rice	Leaf	Heneicosanoic acid, threonic acid, palmitoleic and palmitic acid, non-adecanoic acid and linoleic acid	GC/MS	Agarrwal *et al.* (2014)
Xanthomonas oryzae pv. oryzae	Rice	Leaf	Phenylalanine and tyrosine	GC/TOF and LC/TOF	Sana *et al.* (2010)
Nilaparvata lugens	Rice	Leaf	GABA and glyoxylate	GC/MS	Peng *et al.* (2016)
Fusarium graminearum	Maize	Roots	Metabolites smiglaside and smilaside	LC/MS	Zhou *et al.* (2019)
Bipolaris maydis	Maize	Leaf	Lignin, flavonoids and polyphenols	FT-IR and NMR	Vasmatkar *et al.* (2019)
Pseudomonas syringae	Tomato	Leaf	Flavonoid and phenylpropanoids	NMR and LC/MS	Lopez-Gresa *et al* (2010)

Conclusion

The omics technologies find a vast array of application from understanding the plant stress behavior, specific proteins and metabolites involved in the signaling pathways to utilizing them for improving the genetic gain in plants. Yet, their success would further vest on exploring emerging techniques for their identification, developing and updating databases with the large sets of proteins and metabolites identified for further research. Also, the bioinformatic and statistical tools improvement and integration in the research would increase the proficiency of the techniques. Integrating multiple omics techniques helps better connect among the phenotype and genotype aiding in precise selection of the traits and prepare the future-ready crops.

REFERENCES

Agarrwal, R., Bentur, J. S., and Nair, S. (2014). Gas chromatography mass spectrometry based metabolic profiling reveals biomarkers involved in rice gall midge interactions. *Journal of integrative plant biology, 56*(9), 837-848.

Aghaei, K., Ehsanpour, A. A., and Komastsu, S (2008). Proteome analysis of potato under salt stress. *J. proteome Res.* 7, 4858-4868. Doi: 10.1021/pr800460y

Agrawal, G. K., Sarkar, A., Righetti, P. G., Pedreschi, R., Carpentier, S. and Wang, T., 2013. A decade of plant proteomics and mass spectrometry: translation of technical advancements to food security and safety issues. Mass Spectrom. Rev. 32, 335–365

Alonso, R., Salavert, F., Garcia-Garcia, F., Carbonell-Caballero, J., Bleda, M., Garcia-Alonso, L.,. and Dopazo, J. (2015). Babelomics 5.0: functional interpretation for new generations of genomic data. *Nucleic acids research, 43*(W1), W117-W121.

Boiteau RM, Hoyt DW, Nicora CD *et al.* (2018) Structure elucidation of unknown metabolites in metabolomics by combined NMR and MS/MS prediction. Metabolites 8(1): 8.

Braga, C. P., and Adamec, J. (2018). Metabolome analysis. In *Encyclopedia of Bioinformatics and Computational Biology: ABC of Bioinformatics* (pp. 463-475). Elsevier.

Chebrolu, K. K., Fritschi, F. B., Ye, S., Krishnan, H. B., Smith, J. R., and Gillman, J. D. (2016). Impact of heat stress during seed development on soybean seed metabolome. *Metabolomics, 12*(2), 1-14.

Chun, H., and Kele°, S. (2010). Sparse partial least squares regression for simultaneous dimension reduction and variable selection. *Journal of the Royal Statistical Society: Series B (Statistical Methodology), 72*(1), 3-25.

Coutinho, I. D., Henning, L. M. M., Döpp, S. A., Nepomuceno, A., Moraes, L. A. C., Marcolino-Gomes, J.,. and Colnago, L. A. (2018). Flooded soybean metabolomic analysis reveals important primary and secondary metabolites involved in the hypoxia stress response and tolerance. *Environmental and Experimental Botany, 153,* 176-187.

Davidson, R. L., Weber, R. J., Liu, H., Sharma-Oates, A., and Viant, M. R. (2016). Galaxy-M: A Galaxy workflow for processing and analyzing direct infusion and liquid chromatography mass spectrometry-based metabolomics data. *Gigascience, 5*(1), s13742-016.

Draper, J., Enot, D. P., Parker, D., Beckmann, M., Snowdon, S., Lin, W., and Zubair, H. (2009). Metabolite signal identification in accurate mass metabolomics data with MZedDB, an interactive m/z annotation tool utilising predicted ionisationbehaviour'rules'. *BMC bioinformatics, 10*(1), 1-16.

Fernandez, O., Urrutia, M., Bernillon, S., Giauffret, C., Tardieu, F., Le Gouis, J.,. and Gibon, Y. (2016). Fortune telling: metabolic markers of plant performance. *Metabolomics, 12*(10), 1-14.

Fiehn, O., Barupal, D. K., and Kind, T. (2011). Extending biochemical databases by metabolomic surveys. *Journal of Biological Chemistry, 286*(27), 23637-23643.

Forsberg, E. M., Huan, T., Rinehart, D., Benton, H. P., Warth, B., Hilmers, B., and Siuzdak, G. (2018). Data processing, multi-omic pathway mapping, and metabolite activity analysis using XCMS Online. *Nature protocols, 13*(4), 633-651.

Ghosson, H., Schwarzenberg, A., Jamois, F., and Yvin, J. C. (2018). Simultaneous untargeted and targeted metabolomics profiling of underivatized primary metabolites in sulfur-deficient barley by ultra-high performance liquid chromatography-quadrupole/time-of-flight mass spectrometry. *Plant Methods, 14*(1), 1-17.

Gomes, A. M., Rodrigues, A. P., António, C., Rodrigues, A. M., Leitão, A. E., Batista-Santos, P.,. and Ramalho, J. C. (2020). Drought response of cowpea (Vigna unguiculata (L.) Walp.) landraces at leaf physiological and metabolite profile levels. *Environmental and Experimental Botany, 175,* 104060.

Goufo, P., Moutinho-Pereira, J. M., Jorge, T. F., Correia, C. M., Oliveira, M. R., Rosa, E. A.,. and Trindade, H. (2017). Cowpea (Vigna unguiculata L. Walp.) metabolomics: osmoprotection as a physiological strategy for drought stress resistance and improved yield. *Frontiers in Plant Science, 8,* 586.

Gunnaiah, R., Kushalappa, A. C., Duggavathi, R., Fox, S., and Somers, D. J. (2012). Integrated metabolo-proteomic approach to decipher the mechanisms by which wheat QTL (Fhb1) contributes to resistance against Fusarium graminearum. *PloS one, 7*(7), e40695.

Guo, X., Sarup, P., Jensen, J. D., Orabi, J., Kristensen, N. H., Mulder, F. A.,. and Jensen, J. (2020). Genetic variance of metabolomic features and their relationship with malting quality traits in spring barley. *Frontiers in plant science, 11,* 575467.

He, X. S., Xi, B. D., Li, W. T., Gao, R. T., Zhang, H., Tan, W. B., and Huang, C. H. (2015). Insight into the composition and evolution of compost-derived dissolved organic matter using high-performance liquid chromatography combined with Fourier transform infrared and nuclear magnetic resonance spectra. *Journal of Chromatography A, 1420,* 83-91.

He, Y., and Concheiro Guisan, M. (2019). Microextraction sample preparation techniques in forensic analytical toxicology. *Biomedical Chromatography, 33*(1), e4444.

Herzog, M., Fukao, T., Winkel, A., Konnerup, D., Lamichhane, S., Alpuerto, J. B.,. and Pedersen, O. (2018). Physiology, gene expression, and metabolome of two wheat cultivars with contrasting submergence tolerance. *Plant, Cell and Environment, 41*(7), 1632-1644.

Isaacson, T., and Rose, J. K. (2006). Surveying the plant cell wall proteome, or secretome. *Annual Plant Reviews Volume 28: Plant Proteomics,* 185-209.

Kim, H. K., and Verpoorte, R. (2010). Sample preparation for plant metabolomics. *Phytochemical Analysis: An International Journal of Plant Chemical and Biochemical Techniques, 21*(1), 4-13.

Lewandowska D, ten Have S, Hodge K, Tillemans V, Lamond AI, Brown JW. Plant SILAC: stable-isotope labelling with amino acids of arabidopsis seedlings for quantitative proteomics. PLoS One. 2013 Aug 20;8(8): e72207. doi: 10.1371/journal.pone.0072207. PMID: 23977254; PMCID: PMC3748079.

Li, H., and Tang, H. (2014). An optimized method for NMR-based plant seed metabolomic analysis with maximized polar metabolite extraction efficiency, signal-to-noise ratio, and chemical shift consistency. *Analyst, 139*(7), 1769-1778.

Lin, Y., Ma, C., Liu, C., Wang, Z., Yang, J., Liu, X.,. and Wu, R. (2016). NMR-based fecal metabolomics fingerprinting as predictors of earlier diagnosis in patients with colorectal cancer. *Oncotarget, 7*(20), 29454.

Liu, Q., Wang, X., Tzin, V., Romeis, J., Peng, Y., and Li, Y. (2016). Combined transcriptome and metabolome analyses to understand the dynamic responses of rice plants to attack by the rice stem borer Chilosuppressalis (Lepidoptera: Crambidae). *BMC Plant Biology, 16*(1), 1-17.

Locke, A. M., Barding Jr, G. A., Sathnur, S., Larive, C. K., and Bailey Serres, J. (2018). Rice SUB1A constrains remodelling of the transcriptome and metabolome during submergence to facilitate post submergence recovery. *Plant, Cell and Environment, 41*(4), 721-736.

López Gresa, M. P., Maltese, F., Bellés, J. M., Conejero, V., Kim, H. K., Choi, Y. H., and Verpoorte, R. (2010). Metabolic response of tomato leaves upon different plant–pathogen interactions. *Phytochemical Analysis: An International Journal of Plant Chemical and Biochemical Techniques, 21*(1), 89-94.

Ma, X., Xia, H., Liu, Y., Wei, H., Zheng, X., Song, C.,. and Luo, L. (2016). Transcriptomic and metabolomic studies disclose key metabolism pathways contributing to well-maintained photosynthesis under the drought and the consequent drought-tolerance in rice. *Frontiers in Plant Science, 7,* 1886.

Montenegro-Burke, J. R., Aisporna, A. E., Benton, H. P., Rinehart, D., Fang, M., Huan, T.,. and Siuzdak, G. (2017). Data streaming for metabolomics: accelerating data processing and analysis from days to minutes. *Analytical chemistry, 89*(2), 1254-1259.

Nam, K. H., Kim, D. Y., Kim, H. J., Pack, I. S., Kim, H. J., Chung, Y. S.,. and Kim, C. G. (2019). Global metabolite profiling based on GC–MS and LC–MS/MS analyses in ABF3-overexpressing soybean with enhanced drought tolerance. *Applied Biological Chemistry, 62*(1), 1-9.

Ong,S. and Mann, M. Mass spectrometry-based proteomics turns quantitative. Nature Chemical Biology 2005, 1, 252

Peng, L., Zhao, Y., Wang, H., Song, C., Shangguan, X., Ma, Y.,. and He, G. (2017). Functional study of cytochrome P450 enzymes from the brown planthopper (NilaparvatalugensStål) to analyze its adaptation to BPH-resistant rice. *Frontiers in physiology, 8,* 972.

Ramazi S, Zahiri J. Posttranslational modifications in proteins: resources, tools and prediction methods. Database (Oxford). 2021.

Reich, M., Liefeld, T., Gould, J., Lerner, J., Tamayo, P., and Mesirov, J. P. (2006). GenePattern 2.0. *Nature genetics, 38*(5), 500-501.

Ren, S., Hinzman, A. A., Kang, E. L., Szczesniak, R. D., and Lu, L. J. (2015). Computational and statistical analysis of metabolomics data. *Metabolomics, 11*(6), 1492-1513.

Roessner, U., Wagner, C., Kopka, J., Trethewey, R. N., and Willmitzer, L. (2000). Simultaneous analysis of metabolites in potato tuber by gas chromatography–mass spectrometry. *the plant journal, 23*(1), 131-142.

Rothemund, D.L., Locke, V.L., Liew, A., Thomas, T.M., *et al.,* Depletion of the highly abundant protein albumin from human plasma using the Gradifl ow. Proteomics 2003, 3, 279–287.

Sana, T. R., Fischer, S., Wohlgemuth, G., Katrekar, A., Jung, K. H., Ronald, P. C., and Fiehn, O. (2010). Metabolomic and transcriptomic analysis of the rice response to the bacterial blight pathogen Xanthomonas oryzae pv. oryzae. *Metabolomics, 6,* 451-465.

Scossa, F., Alseekh, S., and Fernie, A. R. (2021). Integrating multi-omics data for crop improvement. *Journal of plant physiology, 257,* 153352.

Sun, X., and Weckwerth, W. (2012). COVAIN: a toolbox for uni-and multivariate statistics, time-series and correlation network analysis and inverse estimation of the differential Jacobian from metabolomics covariance data. *Metabolomics, 8*(1), 81-93.

Sun, Zhiqi, Shengzhen S. Guo, and Reinhard Fässler. "Integrin-mediated mechanotransduction." *Journal of Cell Biology* 215, no. 4 (2016): 445-456.

Tautenhahn, R., Patti, G. J., Rinehart, D., and Siuzdak, G. (2012). XCMS Online: a web-based platform to process untargeted metabolomic data. *Analytical chemistry, 84*(11), 5035-5039.

Thompson, A., Schäfer, J., Kuhn, K., Kienle, S., Schwarz, J., Schmidt, G., Neumann, T. and Hamon, C., 2003. Tandem mass tags: a novel quantification strategy for comparative analysis of complex protein mixtures by MS/MS. *Analytical chemistry, 75*(8), pp.1895-1904.

Trygg, J., and Wold, S. (2002). Orthogonal projections to latent structures (O PLS). *Journal of Chemometrics: A Journal of the Chemometrics Society, 16*(3), 119-128.

Unraveling the metabolite signatures of maize genotypes showing differential response towards southern corn leaf blight by 1H-NMR and FTIR spectroscopy. *Physiological and Molecular Plant Pathology, 108,* 101441.

Villas Bôas, S. G., Mas, S., Åkesson, M., Smedsgaard, J., and Nielsen, J. (2005). Mass spectrometry in metabolome analysis. *Mass spectrometry reviews, 24*(5), 613-646.

Wishart, D. S., Tzur, D., Knox, C., Eisner, R., Guo, A. C., Young, N., and Querengesser, L. (2007). HMDB: the human metabolome database. *Nucleic acids research, 35*(suppl_1), D521-D526.

Xu, S., Lamm, M. E., Rahman, M. A., Zhang, X., Zhu, T., Zhao, Z., and Tang, C. (2018). Renewable atom-efficient polyesters and thermosetting resins derived from high oleic soybean oil. *Green Chemistry, 20*(5), 1106-1113.

Xu, Y., and Goodacre, R. (2012). Multiblock principal component analysis: an efficient tool for analyzing metabolomics data which contain two influential factors. *Metabolomics, 8*(1), 37-51.

Yadav, A. K., Carroll, A. J., Estavillo, G. M., Rebetzke, G. J., and Pogson, B. J. (2019). Wheat drought tolerance in the field is predicted by amino acid responses to glasshouse-imposed drought. *Journal of Experimental Botany, 70*(18), 4931-4948.

Zeng, W., Peng, Y., Zhao, X., Wu, B., Chen, F., Ren, B., Zhuang, Z., Gao, Q. and Ding, Y., 2019. Comparative proteomics analysis of the seedling root response of drought-sensitive and drought-tolerant maize varieties to drought stress. *International Journal of Molecular Sciences, 20*(11), p.2793.

Zhang, T., Creek, D. J., Barrett, M. P., Blackburn, G., and Watson, D. G. (2012). Evaluation of coupling reversed phase, aqueous normal phase, and hydrophilic interaction liquid chromatography with Orbitrap mass spectrometry for metabolomic studies of human urine. *Analytical chemistry, 84*(4), 1994-2001.

Zhou, S., and Wu, C. (2019). Comparative acetylome analysis reveals the potential roles of lysine acetylation for DON biosynthesis in Fusarium graminearum. *BMC genomics, 20*(1), 1-11. Vasmatkar, P., Kaur, K., Pannu, P. P. S., Kaur, G., and Kaur, H. (2019).

Chapter 7

Integration of Omics Technologies in Modern Plant Breeding

Neeraj Kumar[1], Vivek Kumar Kujur[1], Boddu Sangavi[2] and Ayushi Sharma[3]

[1]*Ph.D. Scholar, Department of Genetics and Plant Breeding, IGKV, Raipur, Chhattisgarh*
[2]*Ph.D. Scholar, Department of Genetics and Plant Breeding, Navsari Agricultural University, Gujarat*
[3]*Ph.D. Scholar, Department of Molecular Biology and Biotechnology, GBPUA&T, Pantnagar*

INTRODUCTION

The increasing world population and the impacts of climate change are presenting unparalleled difficulties for the global food production. Addressing these issues requires improving crucial agronomical characteristics beyond just yield, including traits like adaptation, resistance, and nutritional value. This involves adopting both direct and indirect selection methods. The advancement of next-generation high-throughput screening technologies, known as 'omics', offers the potential to accelerate the enhancement of plant traits and the creation of more environmentally friendly crops (Varshney *et al.*, 2021; Castillejo *et al.*, 2023).

With the term "omics" being a derivative of the Greek word "ome" meaning "whole," omics refer to scientific disciplines that study different types of biological molecules constituting complete biological systems(SETAC,2019) These disciplines encompass genomics, transcriptomics, proteomics, metabolomics, and phenomics (Hasin *et al.*, 2017; Khalid *et al.*, 2019).In contemporary molecular biology, the term "omics" specifically denotes a set of methodologies employed for examining an extensive and comprehensive dataset of a specific category or kind of biological molecule within a cell, tissue, organ or whole organism (Zaitlin 2020).

Plant breeding has always been crucial in advancing agriculture to produce enough food for a increasing population. Technological innovations using OMICS approaches have significantly improved its efficiency, resulting in an increased food supply to meet rising demands. Multi-omics investigations of plants have played an essential role in understanding metabolic pathways and the molecular regulators that influence key traits and growth processes in various plant species (Razzaq *et al.*, 2019). Furthermore, the integration of multi-omics data could interpret gene functions and networks better, under various biotic and abiotic stresses (Figure 7.1) (Yang *et al.*, 2021).

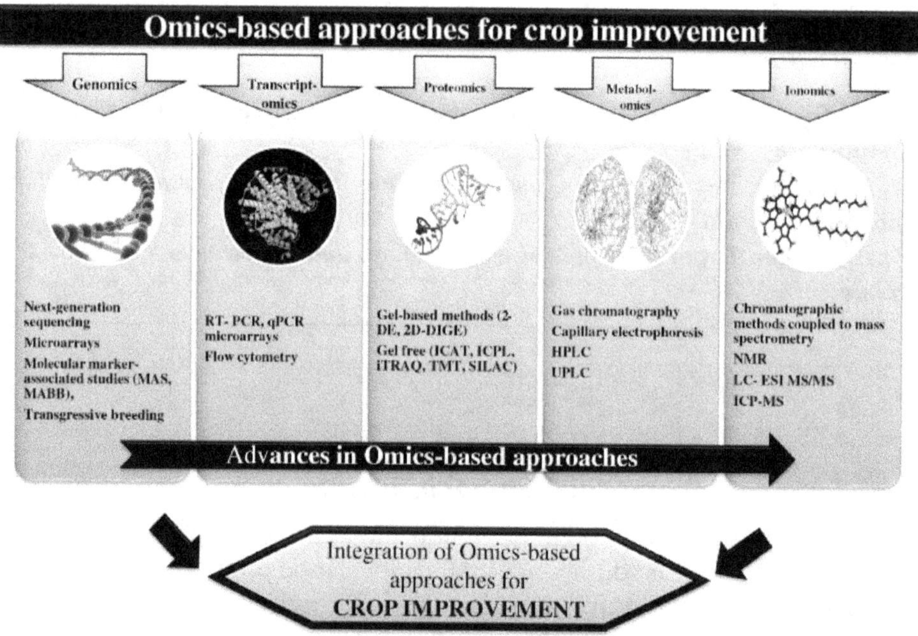

Figure 7.1: Illustration of Diverse Approaches Rooted in Omics that can be Employed to Enhance Crop Quality (Zargar *et al.*, 2016).

Through extensive multi-omics methodologies, scientists have pinpointed critical elements related to senescence, stress reactions, and harvest index in numerous economically significant crops like wheat, rice, soybeans, rapeseed and maize. In this chapter, we illustrate how the multiomics revolution has elevated the effectiveness of plant breeding, which lead to the improvement of nutritional qualities, crop yield and resilience against both biotic and abiotic challenges in crop species, thus promoting sustainable food security. Looking ahead, the integration of multi-omics approaches will have a substantial impact on enhancing genetic advancements and crop breeding techniques.

Harnessing the Power of Proteomic Techniques for Advancing Crop Improvement

The information stored in a plant's DNA are converted into proteins through transcription, specifically mRNA. One of the contemporary approaches to comprehending how genes function on a broader scale involves analyzing the proteins produced by these genes. This field of scientific inquiry is referred to as proteomics. The term "proteome" originates from the concept of the complete set of proteins expressed by a genome. Similar to genomics, "proteomics" pertains to the examination and definition of the entire array of proteins existing within a cell at a specific moment (Wilkins *et al.*, 1995). In contrast to the genome, which is static in nature, the proteome is dynamic. The genome is like a blueprint, while proteome is like the finished product. Post-translational modifications (PTMs) are introduced in the study of proteins and this information is crucial for understanding biological processes.

Proteomics provides novel gene (DNA) identifications to plant scientists and breeders. Damerval *et al.* (1994) used an approach that brought proteomic and MAS components together; they identified protein quantity loci (PQL) that explained some of the spot intensity variation. With help of the analysis of the entire complement of proteins present in crops, proteomics enables the identification of key proteins associated with traits such as stress tolerance, disease resistance, and nutrient utilization. This knowledge guides targeted breeding programs and genetic engineering efforts for enhanced crop performance. For instance, proteomic analysis has been employed to elucidate drought-responsive proteins in rice (Choudhary *et al.*, 2012), highlighting its application in tailoring crops for changing environmental conditions.

Genomics

Genomics is concerned with researching genes and genomes, emphasizing their structure, function, evolution, mapping, epigenomic patterns, mutagenomic influences, and the aspects related to genome editing (Muthamilarasan *et al.*, 2019). It's vital for identifying genetic variation to enhance breeding, significantly improving crops. Structural genomics focuses on genome structure, chromosome organization, and variations, aiding genetic and physical mapping for desired traits. Functional genomics investigates gene roles in these traits. Epigenomic changes like histone modifications or DNA methylation occur genome-wide (Yadav *et al.*, 2018). Mutagenomics studies genetic changes mutants. Pangenomics includes a core and strain-specific genes (Tettelin *et al.*, 2005).

Structural Genomics

Molecular-markers play a crucial role in structural-genomics for mapping and tagging genes linked to desirable characteristics, enhancing crops through breeding. The marker techniques are categorized into non-PCR and PCR-based molecular methods (Muthamilarasan *et al.*, 2019). In the restriction fragment length polymorphism (RFLP) approach, DNA polymorphism is detected through the hybridization of a DNA probe to analyse a restriction enzyme-digested DNA Southern blot, resulting in distinct DNA fragments (Williams *et al.*, 1990).

Randomly amplified polymorphic markers use PCR to amplify short inverted repeats (8-10 bp) in random DNA segments and arbitrarily selected DNA molecule sequences (Rabouam *et al.*, 1999). The amplified fragment length polymorphisms (AFLP) method amplifies genomic restriction fragments through PCR (Vos *et al.*, 1995). Diversity arrays technology utilizes microarray-bound fluorescent DNA probes (Jaccoud *et al.*, 2001). SNPs are a fundamental type of molecular genetic marker, representing unique differences in single nucliotides at specific positions. SNPs, being the most common form of genetic polymorphism, providing dense markers near points of interest. SNP markers have been created for a number of crops, such as rice (Singh *et al.*,2015), sorghum (Ganal *et al.*, 2012) and barley (Bayer *et al.*, 2017).

Two approaches used to understand various characteristics in plants are genome-wide association studies (GWAS) and the mapping of quantitative trait loci (QTL). A statistical technique known as QTL mapping makes it easier to connect different types of information: complex phenotypes (traits with measurements) and genotypes (Challa and Neelapu,2018). GWAS overcomes the limitations of QTL mapping by studying the association between genetic variants, traits and phenotypes present in any organism's population, due to sequence data's SNPs. In sorghum (Lasky *et al.*, 2015) GWAS additionally predicted SNPs linked to drought stress. In response to heat and water stress, GWAS found 48 QTLs for maize crop output (Millet *et al.*, '2016).

Functional Genomics

Functional genomics defines gene functions, aiding transgenics through advanced tools. Identifying interested gene, characterizing them, and developing over expression/knock-out lines are integral (Muthamilarasan *et al.*, 2019). There are three generations of sequencing the genome: first, second and third. Sanger's dideoxy method, in first generation, is manual and tough (Sanger *et al.*, 1977). *Arabidopsis thaliana* pioneered Sanger's way (Arabdopsis Genome Initiative, 2000) Later, it sequenced various additional plants,including *Oryza sativa* (Indica and Japonica) (Yu *et al.*, 2002) and *Glycine max* (Schmutz *et al.*, 2010).

Epigenomics

Epigenomics studies epigenetic changes governing gene regulation in genomes. Methylation of DNA, histone modification and noncoding RNAs (ncRNAs) influence chromatin, activating/suppressing (Yadav *et al.*, 2018). The methylation of DNA studies has been performed to investigated epigenome modifications related to ripening in tomatoes and to examine stable epigenome changes in tissue-cultured rice crops (Stroud *et al.*, 2013).

Mutagenomics

Mutagenomics, a modern omics approach, studies mutational events governing genetic modifications in mutant traits. These can be analyzed using genomics methods like, microarray analysis,TILLING, high-resolution melt (HRM)and serial analysis of gene expression (SAGE) (Penna and Jain, 2017). TILLING in functional genomics characterizes mutagenesis and providing high-throughput mutations in crops. It's successfully applied to identifying crops mutationslike maize (Till- *et al.*,2004),barley (Caldwell- *et al.*,2004), rice(Suzuki- *et al.*,2008)and wheat (Dong- *et al.*,2009).

Pangenomics

Pangenome includes a species' complete genetic makeup, with two sets of genes: core and dispensable. Everyone has common core genes, while dispensable genes are unique or not. Assembling plant pangenomes is vital for species diversity and crop enhancement (Tettelin *et al.*, 2005). Initially, these assemblies used limited accessions and short-read sequencing. Seen in rice (Schatz- *et al.*, 2014), cotton (Chen- *et al.*, 2020), maize (Haberer *et al.*, 2020) and pigeon pea (Zhao- *et al.*, 2020).

Transcriptomics

The branch of "omics technology" known as "transcriptomics" focuses on how an organism's RNA expression profile varies over time and space (Duque *et al.*, 2013). The transcriptome, in contrast to the genome, is far more dynamic and adapts to changes in environment, nutritional availability, age, and developmental stage (El-Metwally *et al.*, 2014). RNA sequencing, microarray platforms, digital gene expression profiling and serial analysis of gene expression (SAGE) are at present used to perform RNA profiling (Duque *et al.*, 2013). By comparing plants under control and stress, this method helps in identifying the candidate genes that are responsible for phenotypic changes and stress tolerance (Le *et al.*, 2012); prediction of tentative gene function and is providing a better crop productivity (Jogaiah *et al.*, 2013). Transcriptomics is concerned with the transcriptome, which is the complete collection of RNA transcripts formed by an organism's genome in a cell or tissue (Raza *et al.*, 2021). Transcriptome profiling is a dynamic process that has gained popularity for examining how genes can be expressed in response to various stimuli across time (Duque *et al.*, 2013). This method helps in understanding a gene's primary function by allowing the investigator to study the differential expression of genes *in vitro*. Traditional profiling, cDNAs-AFLP, differential display-PCR (DD-PCR) and SSH were initially used to evaluate transcriptome dynamics, but these methods had poor resolution (Nataraja *et al.*, 2017). Later on, the development of reliable methods enabled the use of microarrays, digital gene expression profiling, NGS, RNA sequencing, and SAGE for RNA expression profiling (Duque *et al.*, 2013). In response to drought stress, microarray study has resulted that genes in soybean and barley are expressed differently during the embryonic and reproductive stages, respectively (Guo *et al.*, 2009). The comparative genomic analysis of cultivated crops and their wild counterparts is made possible by transcriptomic studies, allowing the identification of further target genes that are essential for the improvement of breeding processes. Transcriptomics is quickly becoming an essential component of plant science research with the improvement of sequencing technology (Zeng *et al.*, 2022). Next-generation sequencing (NGS), also known as RNA sequencing (RNA-seq), and transcriptome analysis allow for the unbiased, high-throughput detection of all expressed transcripts. Additionally, to facilitate better handle the enormous amount of omics data in the future and the ongoing advancement of sequencing technology, the existing analysis methodologies and fundamental assumptions need to be reevaluated and updated (Zeng *et al.*, 2022).

Metabolomics

Metabolomics is a rapidly developing and promising field of analytical biochemistry that focuses on the quantitative and qualitative analysis of all small metabolites found in an organism. Because it can accurately depict the metabolic profile of that living system, so it can be considered of as the culmination of the "omics" cascade (Weckwerth, 2003). Metabolomics developed as a potent tool in basic and applied science due to its broad range of applications. Metabolomics is used to study, characterize, identify, detect, and quantify the metabolic profile of cells, tissues, and living organisms under the specific environmental conditions, is a rapidly expanding, advanced branch of the omics approach (Parida *et al.*, 2018) in the analyses of endogenous metabolites and the metabolites from external sources, the researchers use both non-targeted and targeted techniques (Kosmides *et al.*, 2013). These metabolites consist of lipids, organic acids, aldehydes, ketones, steroids, vitamins, hormones, amino acids, peptides and even secondary metabolites. When compared to proteomics and transcriptomics, this method reproduces data more accurately (Dos Santos *et al.*, 2017). The elucidation of stress tolerance mechanisms with metabolite profiling in plants has been aided by developments in mass spectrometry with liquid or gas chromatography (LC-MS and GC-MS), high-performance liquid chromatography (HPLC), nuclear magnetic resonance spectroscopy (NMR), direct injection mass spectrometry (DIMS), and other metabolomic techniques (Parida *et al.*, 2018).

This is clear from the application of numerous metabolomics techniques to investigate plants and their interacting environment over the past ten years. By using metabolomics and NGS together, it is possible to estimate an early metabolic network from an organism's genomic sequence (Weckwerth, 2011). To facilitate establish methods for crop development, the information is combined using the genome sequencing approaches (NGS) and the quantification of metabolites (via MS) (Pandey *et al.*, 2016). According to Weckwerth and Fiehn (2002), metabolites can be considered as the byproducts of gene expression that represent the biochemical phenotype of the cell. According to Lindon and Nicholson (2008), metabolomics can help discover the biochemical processes that are essential for the proper operation of genes and can as well be used to understand how proteins are expressed metabolically. To create metabolic profiles for a specific plant sample, separation and analytical procedures are needed because different metabolites have varied chemical as well as the physical properties (Jogaiah *et al.*, 2013). Due to the information that plants create more metabolites than either mammals or microorganisms, metabolomics is particularly crucial in plant systems. Plants' production of secondary metabolites aids in coping with environmental stress. The plant responses to various abiotic stresses are correlated with changes in metabolites, the environmental metabolomics is a promising field in stress physiology (Brunetti *et al.*, 2013). In order to give an integrated portrait of numerous activities extending from the genome to the metabolome with phenotypic traits, metabolomics is frequently used in combination with other omics such as genomics, transcriptomics, and proteomics (Weckwerth, 2011).

Phenomics

Phenomics is the sum of all phenotypes at all levels, ranging from molecule to organ. This phrase also refers to whole-organism research in terms of growth, performance, composition, architecture, data collecting, phenotyping employing high-throughput technologies, and data processing. Gerlai coined the term phenomics in 2002 to describe imaging techniques that enable researchers and scientists to learn about plants at the root or whole-plant levels, as well as the inner working mechanisms of leaves. Infrared imaging, 3D imaging, magnetic resonance imaging, fluorescence imaging, and spectral reflectance are examples of phenomics techniques. (Pasala and Pandey, 2020).

For the development of superior plants, the functional evaluation of certain genes and both forward and reverse genetic analysis, phenotyping is crucial. For phenotyping of numerous distinct lines, including germplasm collection, breeding populations, mapping populations, mutant populations, and under various growth conditions, high-throughput phenotyping is also required. Plant phenotype is a complicated interaction between the genotype and the specific environmental factors that affect how a plant grows and develops.

Forward Phenomics

In order to pick the "best of the best" genotype or elite line, forward phenomics uses phenotyping technologies to identify the most promising genotypes with the most desirable attributes within a vast collection. As a result, it makes it unnecessary to grow plants all the way to maturity in the field and permits quick detection of features at the pre-stage (Kumar *et al.*, 2015). The use of "forward phenomics" makes it possible to screen a lot of plant-filled pots that are transported on a conveyor belt and move through a chamber with automated imaging devices (Tsaftaris and Noutsos, 2009). Numerous other crops, including rice, wheat, sorghum, barley, Brassica, and Arabidopsis, were also the subject of studies for trait-specific pheno typing (Yang *etal.*, 2020). The efficiency of the symbiotic relationship between Bradyrhizobia strains and soybean under drought stress can also be determined by comparing the traditional and phenomics methodologies in these research (Govindasamy *et al.*, 2017).

Reverse Phenomics

Reverse phenomics aids in 'best' variations becoming superior traits/ genes because the established mechanisms are also employed to exploit new techniques (Kumar *et al.*, 2015). This entails translating a trait or gene from its physical level to its biochemical and/or later molecular level. For instance, according to (Kumar *et al.*, 2015), researchers have identified the mechanism underlying drought tolerance, which is then followed by the gene or genes responsible for the tolerance. Reverse phenomics enables in-depth mechanistic understanding of features that have been proven to be desirable once the "best of the best" genotype is determined by employing forward genetics. According to Chen *et al.* (2014) and Li *et al.* (2014), plant phenomics contains a number of high-throughput screens, numerous camera units, non-destructive measurements, quantitative analysis, monitor growth dynamics, and stress assessment linked to genomes, opening up new opportunities.

Ionomics: Ionomics in Identifying Novel Genes

Plant ionomics is defined as the "study of the quantitative complement of low molecular weight molecules presents in cells in a particular physiological and developmental state of the plant. The term "ionome" refers to the total mineral nutrient and trace elemental composition and is used to describe the cellular inorganic elements of plant systems (Salt *et al.*, 2008; satismruti *et al.*, 2013). Ionomics includes the quantitative analysis of an organism's elemental makeup and the identification of changes in mineral composition brought on by specific physiological cues, genetic changes, or developmental circumstances.

Different analytical techniques, such as inductively coupled plasma-mass/ optical emission spectroscopy (ICP-MS/OES), neutron activation analysis (NAA) and X-ray crystallography, have been used to acquire ionomics through high throughput elemental profiling in plants (Salt *et al.*, 2008; Kumar *et al.*, 2015).

Conclusion

Plant biologists have used genomics to dissect critical developmental insights and gene characterization. Currently, a multidimensional omics approach, in which a biological sample may be studied for transcriptomics, proteomics and metabolomics in concurrently, *etc.*, provides plant scientists with a comprehensive understanding of plant metabolism by revisiting metabolic pathways or identifying additional pathways.

The use of these omics techniques to investigate how plants react to various biotic and abiotic stresses has made it possible to identify candidate genes, proteins, and metabolites encompassing many quantitative and quality traits of agronomic importance in major crops. Additionally, machine learning helped with the integration of the GWAS to provide an exhaustive collection of data to predict the phenotypic traits influencing the improvement of quality, quantity, or stress tolerance (Yang *et al.*, 2021).

Advances in omics techniques, such as genomes, transcriptomics, proteomics and metabolomics, have cleared the path for speeding crop breeding to cope with climate change and increase food supply. Optimized omics and phenotypic plasticity platform integration will aid in the creation of biological interpretations for complex crop features when in addition to cutting-edge machine learning algorithms.

REFERENCES

Arabidopsis Genome Initiative. (2000). Analysis of the genome sequence of the flowering plant Arabidopsis thaliana. *Nature*, 408(6814), 796–815.

Bayer, M. M., Rapazote-Flores, P., Ganal, M., Hedley, P. E., Macaulay, M., Plieske, J., *et al.* (2017). Development and evaluation of a barley 50k iSelect SNP array. *Frontiers in plant science*, 8, 1792.

Brunetti, C., George, R. M., Tattini, M., Field, K., and Davey, M. P. (2013). Metabolomics in plant environmental physiology. *Journal of experimental botany*, 64(13), 4011-4020.

Caldwell, D.G., McCallum, N., Shaw, P., Muehlbauer, G. J., Marshall, D. F., and Waugh, R. (2004). A structured mutant population for forward and reverse genetics in barley (*Hordeum vulgare* L.). *Plant Journal*, 40(1), 143–150.

Castillejo, M. A., Pascual, J., Jorrín-Novo, J. V., and Balbuena, T. S. (2023). Proteomics research in forest trees: A 2012-2022 update. *Frontiers in Plant Science*, 14, 1130665.

Challa, S., and Neelapu, N. R. (2018). Genome-wide association studies (GWAS) for abiotic stress tolerance in plants. *Biochemical, Physiological and Molecular Avenues for Combating Abiotic Stress Tolerance in Plants*, 135–150. Academic Press.

Chen, D., Neumann, K., Friedel, S., Kilian, B., Chen, M., Altmann, T., and Klukas, C. (2014). Dissecting the phenotypic components of crop plant growth and drought responses based on high-throughput image analysis. *The plant cell,* 26(12), 4636-4655.

Chen, Z. J., Sreedasyam, A., Ando, A., Song, Q., De Santiago, L.M., Hulse-Kemp, A. M., *et al.* (2020). Genomic diversifications of five Gossypium allopolyploid species and their impact on cotton improvement. *Nature genetics* 52(5), 525–533.

Choudhary, M. K., Basu, D., Datta, A., and Chakraborty, S. (2012). Dehydration-responsive reversible and irreversible changes in the extracellular matrix: Comparative proteomics of chickpea genotypes with contrasting tolerance. *Journal of Proteomics, 77,* 56-71.

Damerval, C., Maurice, A., Josse, J. M., and De Vienne, D. (1994). Quantitative trait loci underlying gene product variation: a novel perspective for analyzing regulation of genome expression. *Genetics,* 137(1), 289-301.

Dong, C., Dalton-Morgan, J., Vincent, K., and Sharp, P. (2009). A modified tilling method for wheat breeding. *Plant Genome,* 2(1), 39–47.

Dos Santos, V. S., Macedo, F. A., Do Vale, J. S., Silva, D. B., and Carollo, C. A. (2017). Metabolomics as a tool for understanding the evolution of Tabebuia sensulato. *Metabolomics,* 13, 1-11.

Duque, A. S., de Almeida, A. M., da Silva, A. B., da Silva, J. M., Farinha, A. P., Santos, D.,*et. al.* (2013). Abiotic stress responses in plants: unraveling the complexity of genes and networks to survive. *Abiotic stress-plant responses and applications in agriculture,* 49-101.

El-Metwally, S., Ouda, O. M., Helmy, M., El-Metwally, S., Ouda, O. M., and Helmy, M. (2014). First-and next-generations sequencing methods. *Next generation sequencing technologies and challenges in sequence assembly,* 29-36.

Ganal, M.W., Polley, A., Graner, E. M., Plieske, J., Wieseke, R., Luerssen, H., *et al.* (2012). Large SNP arrays for genotyping in crop plants. *Journal of Biosciences,* 37(5), 821–828.

Govindasamy, V., George, P., Aher, L., Ramesh, S. V., Thangasamy, A., Anandan, S., *et al.* (2017). Comparative conventional and phenomics approaches to assess symbiotic effectiveness of Bradyrhizobia strains in soybean (*Glycine max* L. Merrill) to drought. *Scientific Reports,* 7(1), 6958.

Guo, P., Baum, M., Grando, S., Ceccarelli, S., Bai, G., Li, R. V. K., *et al.* (2009). Differentially expressed genes between drought-tolerant and drought-sensitive barley genotypes in response to drought stress during the reproductive stage. *Journal of experimental botany,* 60(12), 3531-3544.

Haberer, G., Kamal, N., Bauer, E., Gundlach, H., Fischer, I., Seidel, M. A., *et al.* (2020). European maize genomes highlight intraspecies variation in repeat and gene content. *Nature. Genetcs,* 52(9), 950–957.

Hasin, Y., Seldin, M., and Lusis, A. (2017). Multi-omics approaches to disease. *Genome biology,* 18(1), 1-15.

Jaccoud, D., Peng, K., Feinstein, D., and Kilian, A. (2001). Diversity arrays: A solid state technology for sequence information independent genotyping. *Nucleic Acids Research,* 29(4), e25.

Jogaiah, S., Govind, S. R., and Tran, L. S. P. (2013). Systems biology-based approaches toward understanding drought tolerance in food crops. *Critical reviews in biotechnology,* 33(1), 23-39.

Khalid, N., Aqeel, M., and Noman, A. (2019). System biology of metal tolerance in plants: an integrated view of genomics, transcriptomics, metabolomics, and phenomics. *Plant Metallomics and Functional Omics: A System-Wide Perspective,* 107-144.

Kosmides, A. K., Kamisoglu, K., Calvano, S. E., Corbett, S. A., and Androulakis, I. (2013). Metabolomic fingerprinting: challenges and opportunities. *Critical Reviews™ in Biomedical Engineering,* 41(3).

Kumar, J., Pratap, A., and Kumar, S. (2015). Plant phenomics: an overview. *Phenomics in crop plants: trends, options and limitations,* 1-10.

Lasky, J. R., Upadhyaya, H. D., Ramu, P., Deshpande, S., Hash, C. T., Bonnette, J., *et al.* (2015). Genome-environment associations in sorghum landraces predict adaptive traits. *Science advances,* 1(6), e1400218.

Le, D. T., Nishiyama, R., Watanabe, Y., Tanaka, M., Seki, M., Ham, L. H., *et al.* (2012). Differential gene expression in soybean leaf tissues at late developmental stages under drought stress revealed by genome-wide transcriptome analysis. *PloS one,* 7(11), e49522.

Li, L., Zhang, Q., and Huang, D. (2014). A review of imaging techniques for plant phenotyping. *Sensors,* 14(11), 20078-20111.

Lindon, J. C., and Nicholson, J. K. (2008). Analytical technologies for metabonomics and metabolomics, and multi-omic information recovery. *TrAC Trends in Analytical Chemistry,* 27(3), 194-204.

Millet, E. J., Welcker, C., Kruijer, W., Negro, S., Coupel-Ledru, A., Nicolas, S. D., *et al.* (2016). Genome-wide analysis of yield in Europe: allelic effects vary with drought and heat scenarios. *Plant Physiology,* 172(2), 749–764.

Muthamilarasan, M., Singh, N. K., and Prasad, M. (2019). Multi-omics approaches for strategic improvement of stress tolerance in underutilized crop species: a climate change perspective. *Advances in genetics,* 103, 1–38.

Nataraja, K. N., Madhura, B. G., and Parvathi, M. S. (2017). Omics: Modern tools for precise understanding of drought adaptation in plants. In *Plant omics and crop breeding*, 263-294. Apple Academic Press.

Pandey, M. K., Roorkiwal, M., Singh, V. K., Ramalingam, A., Kudapa, H., Thudi, M., *et al.* (2016). Emerging genomic tools for legume breeding: current status and future prospects. *Frontiers in Plant Science*, 7, 455.

Parida, A. K., Panda, A., and Rangani, J. (2018). Metabolomics-guided elucidation of abiotic stress tolerance mechanisms in plants. In *Plant metabolites and regulation under environmental stress*, 89-131. Academic Press.

Pasala, R., and Pandey, B. B. (2020). Plant phenomics: High-throughput technology for accelerating genomics. *Journal of biosciences*, 45(1), 111.

Penna, S., and Jain, S. M. (2017). Mutant resources and mutagenomics in crop plants. *Emirates Journal. Food Agriculture*, 29, 651–657.

Rabouam, C., Comes, A.M., Bretagnolle, V., Humbert, J. F., Periquet, G., and Bigot, Y. (1999). Features of DNA fragments obtained by random amplified polymorphic DNA (RAPD) assays. *Molecular Ecology*, 8(3), 493–503.

Raza, A., Tabassum, J., Kudapa, H., and Varshney, R. K. (2021). Can omics deliver temperature resilient ready-to-grow crops. *Critical Reviews in Biotechnology*, 41(8), 1209-1232.

Razzaq, A., Sadia, B., Raza, A., Khalid Hameed, M., and Saleem, F. (2019). Metabolomics: A way forward for crop improvement. *Metabolites*, 9(12), 303.

Salt, D. E., Baxter, I., and Lahner, B. (2008). Ionomics and the study of the plant ionome. *Annual Review Plant Biology*, 59, 709-733.

Sanger, F., Nicklen, S., and Coulson, A. R. (1977). DNA sequencing with chain-terminating inhibitors. *Proceedings of the National Academy of Sciences*, 74(12), 5463–5467.

Satismruti, K., Senthil, N., Vellaikumar, S., Ranjani, R. V., and Raveendran, M. (2013). Plant ionomics: a platform for identifying novel gene regulating plant mineral nutrition.

Schatz, M. C., Maron, L.G., Stein, J. C., Hernandez Wences, A., Gurtowski, J., Biggers, E., *et al.* (2014). Whole genome de novo assemblies of three divergent strains of rice, Oryza sativa, document novel gene space of aus and indica. *Genome Biology*, 15, 1-16.

Schmutz, J., Cannon, S. B., Schlueter, J., Ma, J., Mitros, T., Nelson, W., *et al.* (2010). Genome sequence of the palaeopolyploid soybean. *Nature*, 463(7278), 178–183.

SETAC. (2019). Technical Issue Paper: OMICS: Complete Systems and Complete Analyses (Issue No. 4). Pensacola, FL.

Singh, N., Jayaswal, P. K., Panda, K., Mandal, P., Kumar, V., Singh, B., *et al.* (2015). Single copy gene based 50 K SNP chip for genetic studies and molecular breeding in rice. *Scientific Reports*, 5(1), 11600.

Stroud, H., Ding, B., Simon, S.A., Feng, S., Bellizzi, M., Pellegrini, M., *et al.* (2013). Plants regenerated from tissue culture contain stable epigenome changes in rice. *Elife*, 2: e00354.

Suzuki, T., Eiguchi, M., Kumamaru, T., Satoh, H., Matsusaka, H., Moriguchi, K., *et al.* (2008). MNU-induced mutant pools and high-performance TILLING enable finding of any gene mutation in rice. *Molecular Genetics and Genomics*, 279, 213–223.

Tettelin, H., Masignani, V., Cieslewicz, M. J., Donati, C., Medini, D., Ward, N. L., *et al.* (2005). Genome analysis of multiple pathogenic isolates of Streptococcus agalactiae: implications for the microbial "pan-genome". *Proceedings of the National Academy of Sciences*, 102(39), 13950–13955.

Till, B. J., Reynolds, S. H., Weil, C., Springer, N., Burtner, C., Young, K., *et al.* (2004). Discovery of induced point mutations in maize genes by TILLING. *BMC Plant Biology*, 4: 1-8.

Tsaftaris, S. A., and Noutsos, C. (2009). Plant phenotyping with low-cost digital cameras and image analytics. In *Information Technologies in Environmental Engineering* (Berlin, Heidelberg: Springer Berlin Heidelberg), 238-251.

Varshney, R. K., Bohra, A., Roorkiwal, M., Barmukh, R., Cowling, W. A., Chitikineni, A., *et al.* (2021). Fast-forward breeding for a food-secure world. *Trends in Genetics*, 37(12), 1124-1136.

Vos, P., Hogers, R., Bleeker, M., Reijans, M., Lee, T. V. D., Hornes, M., *et al.* (1995). AFLP: a new technique for DNA fingerprinting. *Nucleic acids research*, 23(21), 4407–4414.

Weckwerth, W. (2003). Metabolomics in systems biology. *Annual review of plant biology*, 54(1), 669-689.

Weckwerth, W. (2011). Unpredictability of metabolism–the key role of metabolomics science in combination with next-generation genome sequencing. *Analytical and Bioanalytical Chemistry*, 400, 1967-1978.

Wilkins, M. R., Sanchez, J. C., Gooley, A. A., Appel, R. D., Humphery-Smith, I., Hochstrasser, D. F., and Williams, K. L. (1996). Progress with proteome projects: why all proteins expressed by a genome should be identified and how to do it. *Biotechnology and genetic engineering reviews*, 13(1), 19-50.

Williams, J. G. K., Kubelik, A. R., Livak, K. J., Rafalski, J. A., and Tingey, S. V. (1990). DNA polymorphisms amplified by arbitrary primers are useful as genetic markers. *Nucleic Acids Research*, 18(22), 6531–6535.

Yadav, C. B., Pandey, G., Muthamilarasan, M., and Prasad, M. (2018). Epigenetics and epigenomics of plants. *Plant Genetics and Molecular Biology*, 237-261.

Yang, W., Feng, H., Zhang, X., Zhang, J., Doonan, J. H., Batchelor, W. D., *et al.* (2020). Crop phenomics and high-throughput phenotyping: past decades, current challenges, and future perspectives. *Molecular Plant*, 13(2), 187-214.

Yang, Y., Saand, M. A., Huang, L., Abdelaal, W. B., Zhang, J., Wu, Y.,*et. al.* (2021). Applications of multi-omics technologies for crop improvement. *Frontiers in Plant Science*, 12, 563953.

Yu, J., Hu, S., Wang, J., Wong, G. K., Li, S., Liu, B., *et al.* (2002). A draft sequence of the rice genome (*Oryza sativa* L. ssp. indica). *Science*, 296, 79–92.

Zaitlin, D. (2020). Literature Review on the Use of Biotechnology and Omics.

Zargar, S. M., Gupta, N., Nazir, M., Mir, R. A., Gupta, S. K., Agrawal, G. K., and Rakwal, R. (2016). Omics–a new approach to sustainable production. In *Breeding oilseed crops for sustainable production*, 317-344. Academic Press.

Zeng, Z., Li, Y., Li, Y., and Luo, Y. (2022). Statistical and machine learning methods for spatially resolved transcriptomics data analysis. *Genome biology*, 23(1), 1-23.

Zhou, Q., Tang, D., Huang, W., Yang, Z., Zhang, Y., Hamilton, J.P., *et al.* (2020). Haplotype-resolved genome analyses of a heterozygous diploid potato. *Nature Genetics*, 52(10), 1018–1023.

Chapter 8

Plant Epigenomics: Insights into Epigenetic Modification and their Impacts

Gopi M. Patel, Ashita Patel, Harshita R. Patel, Purnima Ray, Maulik Prajapati and Dhrumil Patel*

Department of Genetics and Plant Breeding, NMCA, Navsari Agricultural University, Navsari – 396 450, Gujarat
**e-mail: gopipatel158@gmail.com*

INTRODUCTION

The study of plant epigenetics is currently a captivating and promising field. Advancements in technology have opened up unprecedented possibilities for investigating chromatin modifications, gene expression, and genome structure. Classic epigenetic phenomena, such as transposable element inactivation, imprinting, paramutation, transgene silencing, and co-suppression, were initially observed in plants. The integration of classical genetic research with the latest sequencing technologies now allows us to study these and other epigenetic events in intricate detail, something that were unimaginable just a few years ago.

Research on epigenetics in plants holds immense significance. Plants heavily rely on changes in gene expression to respond to environmental cues, and chromatin-based regulation likely plays a critical role in orchestrating these responses. Moreover, the degree of chromatin "resetting" during sexual reproduction seems to be lower in plants compared to animals, which raises the possibility of inheriting epimutations acquired during the plant's lifespan. Additionally, many plant species can reproduce asexually, producing vegetative clones that present opportunities for the mitotic inheritance of epigenetic states associated with important traits.

Epialleles, distinct from persistent nucleotide mutations, arise from variations in gene methylation among individuals, leading to unique heritable traits passed down through generations. Epigenetics, on the other hand, deals with the study of heritable variations in gene expression that do not involve changes to the underlying DNA sequence, resulting in a change in phenotype without a change in genotype. The Epigenome represents a record of chemical modifications to the DNA and histone proteins within an organism. Changes in gene expression are influenced by modifications to DNA bases, histone proteins, and non-coding RNA biogenesis. Epigenetic mechanisms are crucial in plant defence responses to various stresses, whether caused by biotic factors (*e.g.*, pathogens, insects) or abiotic factors (*e.g.*, salinity, drought, cold, chemical toxicity, pH, heat). Plants adapt and protect themselves at the molecular level through epigenetic regulation, enhancing their survival and tolerance to different stressors.

These understanding highlights that heritable phenotypic variation can occur independently of DNA variation. The term "Epigenetic," introduced by Conrad Waddington, often regarded as the Father of Epigenetics in 1942, illustrated how a set of DNA sequences can dynamically contribute to the development of a complex organism.

Regulatory Mechanisms Controlling Epigenetics

There are three primary regulatory mechanisms that govern epigenetic processes, namely DNA methylation, histone modification, and RNA-mediated interference.

1. DNA Methylation

DNA methylation is a chemical modification facilitated by cytosine methyltransferase, which involves the addition of a methyl group (-CH3) to specific cytosine residues in a DNA sequence, primarily within CpG dinucleotides. This addition of a methyl group creates 5-methyl cytosine, providing a platform for the attachment of various protein complexes that subsequently modify the histone scaffolds. As a consequence, gene expression is altered. This process occurs at the fifth carbon position of the cytosine ring. Methylation can be categorized into two types based on the target site: symmetrical (involving CG and CHG) and asymmetrical (involving CHH). In plants, cytosine can be methylated at CpG, CHpG and CpHpH sites, with H representing any nucleotide. In summary, plants exhibit both symmetrical and asymmetrical types of DNA methylation.

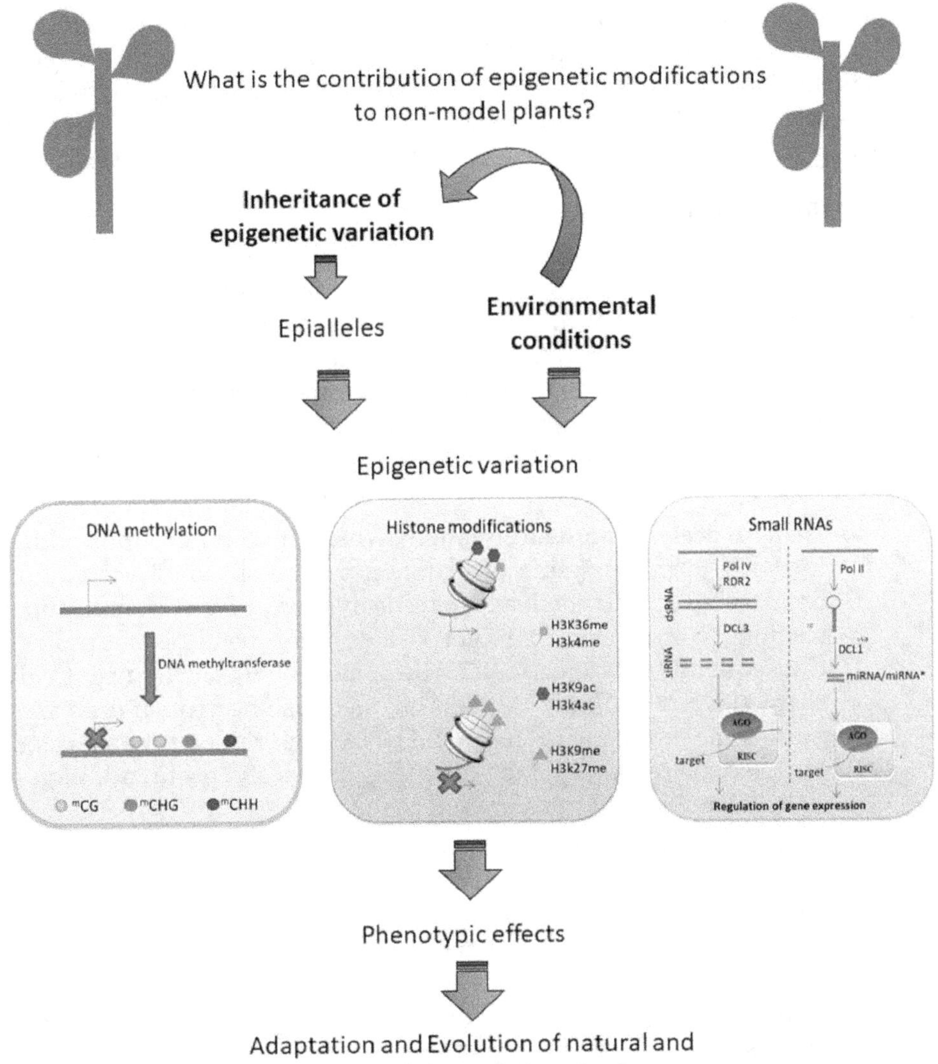

Figure 8.1: Simplified Overview of Main Epigenetic Modifications

(Reproduced from F. Thiebaut., A. S. Hemerly and P. C. G. Ferreira (2019). A Role for Epigenetic Regulation in the Adaptation and Stress Responses of Non-model Plants. *Frontier in plant science.* 10 246-252.)

In plants, specific genes are methylated mainly at CG sites within the gene sequence. However, when it comes to transposable elements (TE), they undergo methylation at CG, CHG, and CHH positions. The consequence of TE and gene promoter region methylation is gene silencing, whereas methylation within the gene leads to gene expression induction. Methylcytosine within a gene directly impacts its expression, whereas in repetitive sequences (TE), methylation serves to safeguard normal genome function. When there is an increase in methylated gDNA, gene expression is reduced, allowing the plant to conserve energy in response to biotic or abiotic stresses. On the other hand, decreasing methylation in resistance-related genes promotes chromatin activation and the expression of novel genes, providing long-term or permanent resistance to stress.

2. Histone Modification

Histone protein modification also plays a significant role in the field of epigenetics. Histones are a group of proteins that serve as structural components of DNA organization in eukaryotes and exist in several variants. In the genome, nucleosomes are formed by two copies of each histone type, namely H2A, H2B, H3, and H4, coming together to form a histone octamer around which 147 base pairs of DNA sequence are wrapped. Chemical alterations, such as acetylation or methylation, affect the N-terminal of histone proteins. These modifications are associated with either the suppression or activation of genes.

a) **Histone acetylation/deacetylation:** When an acetyl group is added to the N-terminal lysine of histones, a process called acetylation occurs. This acetylation leads to the activation of DNA transcription, resulting in a reduction of the net positive charge of the histone protein. Consequently, the electrostatic attraction between the negatively charged DNA and the positively charged histone protein decreases, causing the chromatin to loosen and DNA transcription to be activated. The enzyme responsible for catalyzing this acetyl group addition is called histone acetyltransferase (HATs).

b) **Histone methylation**: Histone methylation involves the methylation of specific amino acids, such as arginine and lysine. Depending on the nature of the methylation on these amino acid residues, it can either activate or deactivate genes. Different types of histone proteins contain distinct arginine and lysine residues that undergo various methylation changes. The methylation can occur once, twice, or thrice on these amino acids. For arginine, it is methylated at one or two sites, while lysine undergoes mono, di, and tri-methylation. Methylation has an impact on the hydrophobicity of the histone, which can alter the interactions between histones and DNA. This modification can also create binding sites for various proteins, restricting the binding of transcription machinery and preventing transcription. The process of methylation is catalyzed by specific enzymes; histone lysine methyltransferases (HKMT) are responsible for lysine methylation, while protein arginine methyltransferases (PRMT) catalyze arginine methylation.

c) **Histone phosphorylation:** Histone phosphorylation, an alteration in histone structure, is facilitated by the enzyme protein kinase, which adds phosphorus molecules to serine, threonine, and tyrosine residues, primarily located in the N-terminal of the histone tail. Phosphorylation increases the negative charge on the histone protein, leading to reduced interactions between DNA and histones, resulting in chromatin de-condensation. On the other hand, dephosphorylation, catalyzed by the enzyme phosphatase, reverses the process by removing phosphorus molecules, causing an increase in positive charge on the histone. This, in turn, leads to chromatin condensation.

RNA Mediated Interference

This epigenetic modification process is commonly referred to as post-transcriptional gene silencing (PTGS). In RdDM (RNA-directed DNA methylation), this modification is considered de novo, meaning that methylation takes place primarily in cytosine residues within the RNA-DNA region with sequence identity. The RdDM pathway specifically targets CHH sequences, as the methylation of these sites at numerous silenced loci relies on RNA-guided de novo methylation. However, RdDM has the capability to methylate all sequences, as symmetrical methylation is perpetuated through each DNA replication, followed by maintenance methylation. MiRNA-directed DNA methylation is a captivating molecular phenomenon that highlights the intricate interplay between microRNAs (miRNAs) and epigenetic alterations in the control of gene activity. MiRNAs, small RNA molecules lacking protein-coding potential, renowned for their role in post-transcriptional gene silencing, have recently been identified as key players in a novel mechanism wherein they guide the initiation of DNA methylation patterns. This process entails enlisting a miRNA-induced silencing complex (miRISC) to specific gene sites. By interacting with the miRISC, these miRNAs steer DNA methyltransferases to precise locations within the genome, leading to the addition of methyl groups to cytosine building blocks within CpG sequences. This miRNA-triggered DNA methylation carries significant ramifications for gene expression, potentially stifling target genes and influencing processes such as cell differentiation, development, and disease advancement. Deciphering the intricate communication between miRNAs and DNA methylation provides insights into the intricate mechanisms overseeing gene control and underscores the intricate nature of epigenetic regulation in shaping biological outcomes.

Methods to Study Epigenetic Modifications

1. DNA Methylation

a. **Bisulfite sequencing:** it is also referred to as bisulphite sequencing, is a technique that involves subjecting DNA to bisulfite treatment prior to standard sequencing procedures. This process enables the identification of methylation patterns within the DNA. DNA methylation is an epigenetic modification that was the first to be identified and remains extensively researched. In animals, it mainly entails the addition of a methyl group to the fifth carbon position of cytosine residues within the CpG dinucleotide. This methylation has been linked to the repression of transcriptional activity. Treating DNA with bisulfite leads to the conversion of cytosine residues to uracil, while 5-methylcytosine residues remain unchanged. As a result, bisulfite-treated DNA exclusively preserves methylated cytosines. Consequently, bisulfite treatment introduces specific modifications in the DNA sequence, which are contingent on the methylation status of individual cytosine residues. This process provides precise, single-nucleotide resolution insights into the methylation pattern of a DNA segment. By analyzing the modified sequence, various techniques can be employed to extract this information. Thus, the primary goal of this analysis revolves around distinguishing single nucleotide polymorphisms (cytosines and thymines) arising from the bisulfite conversion process.

b. **Methylation Sensitive Amplification Polymorphism (MSAP):** Methylation Sensitive Amplification Polymorphism (MSAP) stands as a specialized molecular method that illuminates the DNA methylation patterns present in diverse biological specimens. Stemming from the Amplified Fragment Length Polymorphism (AFLP) approach, MSAP has a dedicated focus on distinguishing alterations in DNA methylation across a range of samples. The procedure encompasses the digestion of genomic DNA using two categories of restriction enzymes: one that is sensitive to methylation and another that is not, followed by the attachment of adapters to facilitate subsequent PCR amplification. By employing a series of targeted amplification stages employing both methylation-sensitive and non-sensitive primers, this technique gives rise to unique banding configurations upon gel electrophoresis. These distinct configurations serve to uncover disparities in DNA methylation amongst the samples, granting researchers the ability to decode epigenetic modifications linked to developmental phases, tissues, or experimental circumstances. Despite lacking the capacity for single-nucleotide precision seen in certain alternative methodologies, MSAP retains its significance as a valuable instrument for scrutinizing the dynamic aspects of DNA methylation and its influence on gene expression and biological mechanisms.

c. **By affinity purification** - This method leverages the specific affinity between methylated DNA and proteins that possess a high affinity for methyl groups, such as Methyl-CpG-binding domain (MBD) proteins. In this process, MBD proteins are immobilized onto a solid support, such as magnetic beads, and then applied to a DNA sample. MBD proteins selectively bind to methylated DNA regions, allowing for the purification of methylated DNA fragments. Unmethylated

DNA remains unbound and is subsequently removed. The isolated methylated DNA can then be subjected to further analyses, such as sequencing or quantitative PCR, to explore the distribution and patterns of DNA methylation. DNA methylation affinity purification offers a targeted approach to studying DNA methylation dynamics and its implications in gene regulation, development, and disease, providing a valuable tool for unraveling the intricacies of epigenetic modifications.

2. Histone Modification

    a. Chromatin immunoprecipitation (ChIP) - Chromatin Immuno-precipitation (ChIP) holds a significant role within the realm of molecular biology, facilitating the exploration of interactions between proteins and DNA within chromatin's context. Widely adopted across diverse research domains, encompassing epigenetics, gene regulation, and chromatin architecture, ChIP serves as a potent instrument for revealing the intricate connections established between proteins and specific DNA sequences. The procedure commences with cell crosslinking to safeguard protein-DNA complexes, followed by chromatin fragmentation and subsequent immunoprecipitation utilizing antibodies tailored to the protein of interest. This sequence of events culminates in the enrichment of chromatin regions that harbor the protein interaction, subsequently amenable to scrutiny through methodologies like PCR or sequencing. ChIP empowers researchers to decode the functions played by transcription factors, histone modifications, and regulatory proteins in the orchestration of gene expression. Additionally, it unravels the dynamic aspects of epigenetic changes. ChIP's capacity to furnish insights into the functional implications of protein-DNA interactions positions it as an indispensable approach, pivotal for unveiling the intricate molecular underpinnings governing gene regulation and epigenetic phenomena.

    b. ChiP-Seq - Progress in the realm of next-generation sequencing has facilitated the fusion of Chromatin Immunoprecipitation (ChiP) with cutting-edge sequencing techniques like Solexa. This innovative approach, known as ChIP-Seq, enables the comprehensive exploration of the genomic distribution of histone proteins on a genome-wide scale.

    c. ChiP-PCR- The DNA that has been extracted through immuno-precipitation is amplified and measured using real-time PCR (RT-PCR) techniques employing TaqMan or Syber Green Technologies, along with designated primers. This allows for the examination of distinct genomic regions linked to specific histones.

3. Non-Coding RNA

a. Deep sequencing - Epigenetic deep sequencing, a powerful and sophisticated molecular technique, offers an intricate window into the complex landscape of epigenetic modifications on a genome-wide scale. By integrating the capabilities of deep sequencing technology with the study of epigenetic marks, such as DNA methylation and histone modifications, this approach delivers an unprecedented level of detail in unraveling the regulatory intricacies of gene expression and cellular dynamics. Beginning with the isolation of DNA or RNA from biological samples, epigenetic deep sequencing involves fragmenting and sequencing these molecules in parallel, producing an expansive dataset that captures the spatial distribution and frequency of epigenetic modifications throughout the genome. Leveraging specialized bioinformatics analyses, researchers decode this wealth of information to discern patterns of DNA methylation, histone modifications, and other epigenetic events. This technique finds applications in diverse realms, including unraveling the epigenetic basis of diseases, tracing developmental trajectories, investigating environmental influences, and identifying novel therapeutic targets. Epigenetic deep sequencing serves as a transformative tool, enabling profound insights into the dynamic interplay between epigenetic modifications and the intricate orchestration of biological processes.

Different functions regulated by the epigenetic mechanism

Epigenetic modifications also play a significant role in regulating various functions in plants. Here are some examples:

1. **Flowering Time Regulation:** Epigenetic modifications, such as DNA methylation and histone modifications, influence the timing of flowering in plants. Changes in epigenetic marks can lead to alterations in the expression of genes involved in flowering pathways.

2. **Stress Responses:** Plants can undergo epigenetic changes in response to environmental stresses, such as drought, heat, and pathogens. These changes can affect the expression of stress-responsive genes and enhance the plant's ability to adapt to challenging conditions.

3. **Seed Development and Germination:** Epigenetic modifications are crucial for regulating seed development and germination. Imprinting of specific genes can affect seed size, dormancy, and germination rates.

4. **Nutrient Uptake and Homeostasis:** Epigenetic mechanisms help plants adjust their nutrient uptake and utilization in response to changing soil conditions. These modifications impact the expression of genes involved in nutrient transport and storage.

5. **Somatic Embryogenesis and Regeneration:** Epigenetic modifications play a role in somatic embryogenesis, a process by which plant cells can be induced to develop into embryos without fertilization. Epigenetic changes are also involved in plant regeneration from tissue culture.

6. **Vernalisation:** Vernalisation is the process by which some plants require a period of cold temperature exposure to initiate flowering. Epigenetic modifications, particularly changes in DNA methylation, are involved in this phenomenon.

7. **Pathogen Defense:** Epigenetic changes can contribute to plant defense mechanisms against pathogens. Genes involved in disease resistance are often regulated by epigenetic modifications to enhance the plant's ability to fend off infections.

8. **Epigenetic Inheritance:** Plants can transmit epigenetic information to their offspring. Epigenetic modifications acquired during the parent plant's growth can influence traits in the next generation, even without changes in the DNA sequence.

9. **Hormone Signaling:** Epigenetic modifications are integrated with hormone signaling pathways to regulate plant growth, development, and responses to external cues.

10. **Fruit Ripening:** Epigenetic regulation is involved in fruit ripening processes, influencing changes in color, texture, and flavor as fruits mature.

Conclusion

The epigenome encompasses all epigenetic modifications in DNA bases (while maintaining the original nucleotide sequence), histone proteins, and the biogenesis of small RNAs within a cell. Genome-wide alterations in epigenetic patterns have been observed during cell growth, development, and environmental stress, often linked to changes in gene expression. Following the cessation of stress, gene expression and epigenetic modifications can revert to their pre-stress states. A prominent mechanism of epigenetic alteration involves the methylation of cytosine at the 5th carbon position. Furthermore, specific amino acids in histone proteins undergo post-translational modifications that can impact processes like transcription, chromosome condensation, segregation, and DNA repair. Small RNAs are essential players in DNA methylation through the RNA-directed DNA methylation (RdDM) pathway. Epigenetic changes can be inherited across generations, leading to variations in phenotype. It is increasingly clear that these changes contribute significantly to acclimatization, stress tolerance, adaptation, and evolution. Given the mounting evidence of their influence on gene expression, delving into the epigenetic machinery of gene regulation in plants and exploring its potential for epigenome engineering/ editing becomes crucial, especially for enhancing crops. This concise review provides an overview of epigenomics, discussing its current state and future potential for advancing crop improvement to address the challenges of sustainable global food security.

REFERENCES

Bewick, A. J.; Ji, L.; Niederhuth, C. E.; Willing, E. M.; Hofmeister, B. T. and Shi, X. (2016). On the origin and evolutionary consequences of gene body DNA methylation. *Proceedings of the National Academy of Sciences of the United States of America*, **113**: 9111.

Bossdorf, O.; Richards, C. L. and Pigliucci, M. (2008). *Epigenetics for ecologists. Ecology Letters*, **11**: 106-115

Fuchs, J.; Demidov, D.; Houben, A. and Schubert, I. (2006). Chromosomal histone modification patterns-From conservation to diversity. *Trends in Plant Science*, **11**: 199- 208.

He, G.; Elling, A. A. and Deng, X. W. (2011). *The epigenome and plant development. Annual Review of Plant Biology*, **62**: 411-435.

Hirayama, T. and Shinozaki, K. (2010). Research on plant abiotic stress responses in the postgenome era: Past, present and future. *Plant Journal*, **61**(6): 1041-1052.

Kalisz, S. and Purugganan, M. D. (2004). Epialleles via DNA methylation: consequences for plant evolution. *TRENDS in Ecology and Evolution*, **19**(6): 309-314.

Kanno, T.; Bucher, E.; Daxinger, L.; Huettel, B.; Kreil, D. P. and Breinig, F. (2010). RNAdirected DNA methylation and plant development require an IWR1-type transcription factor. *EMBO Reports*, **11**: 65-71.

Matzke, M.; Kanno, T.; Daxinger, L.; Huettel, B. and Matzke, A. J. (2009). RNA-mediated chromatin-based silencing in plants. *Current Opinion in Cell Biology*, **21**: 367-376.

Miura, A.; Yonebayashi, S.; Watanabe, K.; Toyama, T.; Shimada, H. and Kakutani, T. (2001). Mobilization of transposons by a mutation abolishing full DNA methylation in Arabidopsis. *Nature*, **411**: 212-214.

Niederhuth, C. E. and Schmitz, R. J. (2017). Putting DNA methylation in context: From genomes to gene expression in plants. *Biochimica et Biophysica Acta*, **1860**: 149-156.

Richards, E. J. (2006). Inherited epigenetic variation - Revisiting soft inheritance. *Nature Reviews Genetics*, **7**: 395-401.

Roudier, F.; Ahmed, I.; Berard, C.; Sarazin, A.; Mary-Huard, T. and Cortijo, S. (2011). Integrative epigenomic mapping defines four main chromatin states in Arabidopsis. *The EMBO Journal*, **30**: 1928-1938.

Shahbazian, M. D. and Grunstein, M. (2007). Functions of site-specific histone acetylation and deacetylation. *Annual Review of Biochemistry,* **76**: 75-100.

Simon, W.; Henderson, I. and Jacobsen, S. (2005). Gardening the genome: DNA methylation in Arabidopsis thaliana. *Nature Reviews Genetics,* **6**: 351-360.

Singroha, G. and Sharma, P. (2019). Epigenetic modifications in plants under abiotic stress. *Epigenetics,* **55**(4): 35-41.

Chapter 9

Non-Coding RNAs and its Importance in Crop Improvement

Mandava Harini[1] and Ragulakollu Sravanthi[2]*

[1]Genetics and Plant Breeding, Indira Gandhi Krishi Vishwavidhyalaya, Raipur, Chhattisgarh
[2]Ph.D. Scholar, Genetics and Plant Breeding, Tamil Nadu Agricultural University, Coimbatore, Tamil Nadu
**e-mail: harinimandava99@gmail.com*

INTRODUCTION

Agricultural crops face a diverse array of environmental stresses, which diminish and restrict their productivity. Plants encounter two categories of environmental stresses: abiotic stress and biotic stress. Abiotic stress refers to non-living factors, such as extreme temperatures, water scarcity, and soil degradation, while biotic stress is caused by living organisms like pests, diseases, and invasive plants that adversely affect plant growth and yield (Gull *et al.*, 2019). Additionally, the escalating global population, rapid urbanization leading to diminished arable land, notable alterations in soil quality, and the impacts of global warming are all foreseen to aggravate food security concerns in the future (Mitter, 2006).

Plant cells exhibit varying chromatin states in response to environmental stimuli, enabling them to finely adjust their transcriptional profiles, either temporarily or permanently, to better adapt to their surroundings. Epigenetic modifications, including post-transcriptional histone modifications, histone variants, DNA methylation, and the influence of non-coding RNAs, encompass a diverse array of changes to chromatin states. These modifications can epigenetically govern specific transcriptional outcomes (Varotto *et al.,* 2020). Transcriptional or posttranslational gene regulation assumes a vital role in establishing cellular homeostasis during various stress conditions (Ahmed *et al.,* 2020). Non-coding RNAs(ncRNAs) are a special class of RNAs that do not get translated to form a functional protein. A wide range of ncRNAs exists, serving diverse functions in both prokaryotes and eukaryotes (Peedicayil, 2021).

The DNA regions encoding ncRNAs exhibit similar chromatin modifications to protein-coding genes. The main distinction lies in their coding potential. The essential aspect of ncRNAs is their ability to form secondary RNA structures, which plays a crucial role in controlling the creation and propagation of heterochromatin domains at both neighbouring (cis-acting) and distant (trans-acting) genomic loci.

Non-coding RNAs (ncRNAs) play a crucial role in regulating gene expression and can be categorized into different classes based on their mechanisms of action and genomic origin, encompassing both housekeeping ncRNAs and regulatory ncRNAs (Ariel *et al.,* 2015). The housekeeping ncRNAs comprise of ribosomal RNAs (rRNAs), transfer RNAs, small nuclear RNA (snRNA) and small nucleolar RNAs (snoRNAs). The regulatory ncRNAs are divided into sub-classes: (a) small ncRNAs (sncRNAs) including small interfering RNAs (siRNAs) and microRNAs (miRNAs), and (b) long non-coding RNAs (lncRNAs). lncRNAs can be broadly divided into different classes such as intronic ncRNAs (incRNAs) and long intergenic ncRNAs (lincRNAs) (Ahmed *et al.,* 2020). Recent findings indicate that non-coding RNAs (ncRNAs), particularly microRNAs (miRNAs) and long ncRNAs (lncRNAs), have a substantial impact on the regulation of gene expression in response to VARIOUS PLANT STRESSES (AHMED *et al.,* 2020).

House Keeping ncRNA

Typically, small, ranging from 50 to 500 nucleotides (nt), these ncRNAs are consistently expressed in all cell types and essential for maintaining cell viability. In addition to their vital functions, such as rRNAs and tRNAs in protein synthesis, snRNAs in RNA splicing, and snoRNAs in RNA modifications, certain housekeeping ncRNAs can also serve regulatory roles through cleavage. (Zhang *et al.,* 2019).

Regulatory ncRNA

Regulatory non-coding RNAs are transcribed from DNA; however, they lack the capacity to undergo translation into proteins (Hangauer *et al.,* 2013). They carry out a multitude of essential roles in aspects such as plant growth, development, and responses to abiotic stresses, both during transcription and in post-transcriptional processes. The spectrum of regulatory non-coding RNAs encompasses microRNAs (miRNAs), short interfering RNAs (siRNAs), and long non-coding RNAs (lncRNAs) by (Sang and Chanseok, 2016).

1. miRNA

MicroRNAs are short, single-stranded RNA molecules of approximately 22 nucleotides in length. They are transcribed from specific genomic loci as primary miRNA transcripts (pri-miRNAs) by RNA polymerase II. Pri-miRNAs are processed by the Microprocessor complex, which includes the RNase III enzyme Drosha, to form precursor miRNAs (pre-miRNAs) with hairpin structures. Pre-miRNAs are then exported from the nucleus to the cytoplasm by Exportin-5.

In the cytoplasm, pre-miRNAs are further processed by another RNase III enzyme, Dicer, which cleaves the hairpin structure to generate mature miRNA duplexes. One strand of the duplex, the guide strand, is incorporated into the RNA-induced silencing complex (RISC), while the other strand, the passenger strand, is usually degraded.

Role in Post-Transcriptional Gene Regulation

Once incorporated into the RISC complex, mature miRNAs guide the complex to target mRNAs with complementary sequences. This binding can lead to either mRNA degradation or translational repression, depending on the degree of complementarity between the miRNA and the target mRNA.

Examples of Well-Studied miRNAs in Crop Species:

- ☆ **miR156:** Controls the transition from the vegetative to the reproductive phase in plants by targeting transcription factors.
- ☆ **miR159:** Regulates floral organ identity and development by targeting MYB transcription factors.
- ☆ **miR167:** Modulates root development by targeting auxin response factors.
- ☆ **miR172:** Regulates flowering time and floral organ identity by targeting APETALA2-like transcription factors.

2. siRNA

siRNA is small regulatory ncRNA with a length of around 20-30 nucleotides which are categorized as exogenous and endogenous. When the plants are infected by virus, RNA derived from viral RNA in such plants is called exogenous siRNA. By utilising RNA-dependent RNA polymerase, these abnormal RNA strands form dsRNAs. RNAs (dsRNAs), resulting in the production of primary vsiRNAs measuring 22, 24, and 21 nucleotides. These primary vsiRNAs are then further multiplied through RNA-dependent RNA polymerase (RDR) activity and incorporated into AGO proteins, forming RNA-induced silencing complexes (RISCs). The types of endogenous siRNAs are phased small interfering RNAs (phasiRNAs), heterochromatic siRNAs (het-siRNAs), trans-acting siRNAs (tasiRNAs), and natural antisense siRNAs (natsiRNAs). These diverse siRNAs can arise from internal genetic sequences such as transposons, repetitive elements, or tandem repeats within plants. Their primary function involves the cleavage of target RNAs (Jiang *et al.*, 2020; Garcia-Ruiz *et al.*, 2010; Leonetti *et al.*, 2020).

Role in RNA Interference (RNAi) and Gene Silencing

Once incorporated into the RNA-induced silencing complex (RISC), siRNAs guide the complex to target RNAs with complementary sequences. Unlike miRNAs, siRNAs typically exhibit perfect complementarity to their target sequences. This leads to target mRNA cleavage and degradation, effectively silencing the expression of the target gene.

Applications in Suppressing Pathogen Genes and Improving Stress Tolerance

SiRNAs have significant potential in crop improvement by providing a means to silence specific genes. In agriculture, siRNAs can be used to suppress the expression of genes in pathogens, thus reducing their virulence. Moreover, siRNAs can also enhance stress tolerance in crops by silencing genes that negatively affect stress responses, making plants more resilient to various environmental challenges.

3. lincRNA

Long intergenic non-coding RNAs (lincRNAs) constitute a category of non-coding RNAs exceeding 200 nucleotides in length. These sequences exhibit limited conservation across various species and play a significant role in regulating plant growth and responding to stress conditions (Mercer *et al.*, 2009). The process of their formation mirrors mRNA synthesis and necessitates the involvement of POLII to commence enzymatic transcription, facilitated by transcription factors (TFs). Additionally, certain lincRNAs may undergo transcription by RNA polymerase III, while in plants, a small number of lincRNAs are generated by the plant-specific POLIV and V enzymes (Ponting *et al.*, 2009).

In contrast to mRNAs derived from their respective genes, the sites of lincRNA transcription exhibit variability. Consequently, based on their genomic location relative to neighbouring protein-coding genes, lincRNAs can be categorized into sense lincRNAs and antisense lincRNAs. These two categories align with or complement one (or more) exons of protein-coding genes, occupying overlapping positions. Simultaneously expressed with their corresponding transcripts on opposing strands, bidirectional lincRNAs display a distinctive pattern. Intronic lincRNAs, on the other hand, originate from introns within protein-coding genes. Meanwhile, large intergenic lincRNAs and enhancer lincRNAs are derived from the enhancer regions associated with protein-coding genes.

The transcriptional positioning of lincRNAs across various genomic sites implies a mechanistic framework for lincRNA-mediated gene expression regulation (Zhang *et al.*, 2023).

a. DECOYS

The core regulatory mechanism of transcriptional lincRNAs involves their interaction with RNA-binding proteins (RBPs), constituting the primary means of gene expression regulation. These RBPs that lincRNAs bind to can encompass transcription factors (TFs), chromatin modifying factors, or other regulatory molecules. It's worth noting that lincRNAs are not involved in any other functions. LincRNAs adhering to this regulatory framework not only associate with RBPs essential for typical mRNA transcription, but also exert a repressive influence within the promoter region. This is achieved by forming complexes with RBPs, impeding the usual mRNA transcription process. Eliminating the lincRNA can lead to a restored phenotype.

However, certain lincRNAs in plants operate as decoys and have the opposite effect of enhancing gene expression. These lincRNAs are typically found in intergenic regions, contain multiple coding genes, exhibit lengthy sequences akin to mRNAs, and post-transcriptionally possess multiple miRNA-binding sites. Consequently, these lincRNAs are also termed endogenous competing RNAs.

b. GUIDES

LincRNAs exert control over gene expression through a twofold approach: they recruit proteins directly to bind at specific sites or they establish complexes with proteins that subsequently bind to target gene-specific locations. These target sites can encompass neighbouring genes or genes at a distance (Hung and Chang, 2010). Foretelling this mode of regulation based solely on the lincRNA sequence proves challenging. However, lincRNAs enable modifications in local chromosome structure to yield both local (cis) and remote (trans) effects. Irrespective of whether the influence is cis or trans, the fundamental principle remains constant: the manipulation of chromosomal structural states for the transmission of regulatory cues that govern gene expression. This, in turn, leads to alterations in the epigenome. Notably, lincRNAs such as AIR and eRNAs, which arise from promoters or enhancers, demonstrate this regulatory capacity in cis, affecting gene expression.

c. SCAFFOLDS

Historically, scaffold complexes, comprised of proteins, have been conventionally regarded as sites of assembly and integral constituents within the realm of biological macromolecules. More recent investigations, however, have unveiled an additional role for lincRNAs as pivotal frameworks for the assembly of associated molecules. This newfound function is crucial for the precise regulation of intermolecular interactions and the specificity of signal transduction. By leveraging lincRNAs as scaffolds, disparate or identical proteins can aggregate, giving rise to ribonucleic protein complexes known as lincRNA-RNPs. These lincRNA-RNPs exert influence over target gene histone modifications, functioning within the cis (or trans) context. This is achieved through the binding of transcription factors (TFs), which in turn activate or inhibit gene expression by enlisting the assistance of chromatin modification enzymes (Wang and Chang, 2011).

d. SIGNALS

The expression of genes is notably specific in terms of both spatial and temporal dimensions. In response to varying environmental cues, plants activate distinct genes in different locations and periods. Similarly, the transcription of lincRNAs also adheres to specific timeframes and locales, closely intertwined with the transcription of mRNAs (Wang *et al.*, 2017). Consequently, lincRNAs can function as molecular cues. When transcription factors (TFs) recognize lincRNAs, the ensuing interaction with associated proteins and regulatory factors elicits either positive or negative control over gene expression. This orchestration streamlines the identification of chromatin states within regulatory elements. Beyond this, exploiting RNA as a signaling conduit for gene regulation can accelerate the development of stress resilience in plants. This allows for swift adaptation to environmental pressures, bolstering their adeptness in thriving within their surroundings. Notably, this regulation operates independently of the recognition and catalytic activities of proteins generated through transcription.

Roles of Non-Coding RNAs in Crop Improvement

Non-coding RNAs (ncRNAs) have emerged as key players in create plant traits and responses to environmental challenges. They influence various biological processes, and their manipulation holds significant potential for crop improvement (Sailaja Bhogireddy *et al.*, 2021). The roles of ncRNAs in the context of crop enhancement can be categorized into several important areas:

1. Regulation of Developmental Processes

- ☆ **Flowering Time:** NcRNAs, particularly miRNAs and lncRNAs, regulate the transition from vegetative to reproductive phases. They target genes involved in flowering pathways, ensuring appropriate flowering under varying environmental conditions.

- ☆ **Root Development:** NcRNAs modulate the growth and architecture of roots by regulating genes related to root development, nutrient uptake, and symbiotic interactions with beneficial soil microbes.

- ☆ **Fruit Ripening:** MiRNAs and lncRNAs play a role in controlling the timing and progression of fruit ripening by influencing hormone signaling pathways and the expression of ripening-associated genes.

2. Enhancement of Abiotic Stress Tolerance

☆ **Drought Stress:** NcRNAs are involved in adjusting plant responses to water deficit by regulating genes associated with water transport, stomatal regulation, and osmotic stress tolerance. MiRNAs can fine-tune these processes to optimize water use efficiency.

☆ **Salinity Stress:** NcRNAs help plants cope with high salt levels by modulating ion homeostasis, ion transport, and osmotic adjustment. This allows crops to maintain cellular integrity and function under saline conditions.

☆ **Temperature Stress Responses:** NcRNAs regulate temperature-responsive genes, contributing to the acclimation and adaptation of plants to temperature fluctuations. They modulate heat shock protein expression and other stress-related pathways.

3. Augmentation of Biotic Stress Resistance

☆ **Plant-Pathogen Interactions:** NcRNAs participate in plant immunity by targeting pathogen effectors or manipulating host genes to enhance resistance. They regulate the expression of immune-related genes, contributing to the activation of defense responses.

☆ **Pest Resistance:** NcRNAs play a role in plant resistance against insect pests by regulating genes associated with the production of defense compounds, such as secondary metabolites and toxins.

4. Improvement of Yield and Nutrient Content in Crops

☆ **Yield Enhancement:** NcRNAs can influence crop yield by regulating genes involved in growth, development, and reproductive processes. By targeting yield-related pathways, they offer opportunities for increasing crop productivity.

☆ **Nutrient Content:** NcRNAs can be used to enhance the nutritional quality of crops by modifying the expression of genes involved in nutrient uptake, assimilation, and storage. For instance, manipulating miRNAs can enhance the accumulation of essential nutrients like iron, zinc, and vitamins.

☆ **Drought-Resistant Maize:** Using engineered miRNAs targeting drought-responsive genes, researchers have developed maize varieties with improved water-use efficiency and enhanced drought tolerance, ensuring more stable yields under water scarcity.

☆ **Salt-Tolerant Rice:** By manipulating lncRNAs involved in salt stress signaling, rice varieties with enhanced salt tolerance have been developed. These varieties maintain better ion homeostasis and growth under saline conditions.

☆ **Virus-Resistant Tomatoes:** Small interfering RNAs (siRNAs) targeting viral genes have been used to engineer tomatoes resistant to devastating plant viruses, reducing yield losses and the need for chemical treatments.

Techniques for Studying Non-Coding RNAs

1. **RNA Interference (RNAi) and Gene Silencing:** Design and Delivery of Synthetic miRNAs and siRNAs for Targeted Gene Silencing: Engineered miRNAs and siRNAs can be designed to target specific genes of interest (Santhosh *et al.*, 2015). These synthetic ncRNAs are introduced into plants to induce the degradation of target mRNAs or inhibit their translation (Hua *et al.*, 2019). This approach offers a precise way to downregulate undesirable traits or enhance desired traits without introducing foreign DNA.

2. **Enhancing Desired Traits:** RNA interference (RNAi) technology can be employed to silence genes responsible for undesirable traits such as susceptibility to diseases, pests, or unfavourable environmental conditions. Conversely, RNAi can also be used to enhance traits such as stress tolerance, nutritional content, or yield. For instance, silencing a gene involved in ethylene production could extend the shelf life of harvested fruits.

3. **CRISPR-Cas9 and Genome Editing:** Utilizing ncRNAs for Guiding CRISPR-Cas9 to Specific Genomic Targets: CRISPR-Cas9 is a revolutionary genome editing tool that can be guided to specific DNA sequences using synthetic single-guide RNAs (sgRNAs) (Diriba Guta *et al.*, 2015). These sgRNAs can be engineered from ncRNA scaffolds, and their binding to target DNA triggers the Cas9 enzyme to induce precise double-strand breaks, allowing for gene knockout, knockout, or modification (Hua *et al.*, 2019).

4. **Precision Genome Editing and Regulation of Endogenous Genes:** By leveraging ncRNAs in the CRISPR-Cas9 system, researchers can achieve precise genome editing, such as introducing point mutations, gene deletions, or insertions. Additionally, engineered ncRNAs can guide Cas9 to target specific regulatory elements, influencing gene expression levels without altering the underlying DNA sequence (He J *et al.*, 2019).

5. **Epigenetic Modifications:** Use of ncRNAs to Induce Epigenetic Changes for Stable Trait Alterations: Epigenetic modifications, such as DNA methylation and histone modifications, play a role in regulating gene expression (Shin *et al.*, 2016). Engineered lncRNAs can guide epigenetic-modifying complexes to specific genomic loci, inducing stable changes in gene expression patterns. This approach allows for heritable trait alterations without altering the DNA sequence

Epigenetic changes induced by engineered ncRNAs can be passed down to subsequent generations, leading to transgenerational effects on trait expression. This provides a unique way to introduce and maintain desired traits in crop populations over time (Sang and Chanseok 2016). By harnessing the power of RNA interference, CRISPR-Cas9 genome editing, and epigenetic modifications, researchers can precisely manipulate plant genomes to enhance desirable traits and improve crop performance. These approaches offer new avenues for sustainable crop improvement and agricultural innovation.

Future Perspectives and Challenges

Potential Future Developments in ncRNA Research for Crop Improvement (Sailaja Bhogireddy *et al.*, 2021):

☆ **Advancements in Understanding ncRNA Functions:** As research on ncRNAs continues, we can expect to uncover more intricate roles and mechanisms of various ncRNA classes. This deeper understanding will provide new insights into their potential applications in crop improvement.

☆ **Trait Enhancement:** With advancements in genome editing and ncRNA manipulation techniques, the ability to tailor crops for specific traits will become more refined. This could lead to crops that are better suited for specific environmental conditions, nutritional needs, or consumer preferences.

☆ **Multi-Gene Regulation:** Future research may focus on engineering complex regulatory networks involving multiple ncRNAs to achieve precise and coordinated control over gene expression. This could lead to enhanced trait combinations and more sophisticated crop phenotypes.

Exploration of Emerging ncRNA-Based Technologies and Their Integration into Breeding Programs:

1. **Nanotechnology and ncRNA Delivery:** Nanoparticle-based delivery systems could revolutionize the targeted delivery of ncRNAs into plant cells (Khraiwesh *et al.*, 2012). These systems offer enhanced stability, protection, and controlled release of ncRNAs, increasing the efficiency of their uptake and activity.

2. **Designer ncRNAs:** Computational approaches could be used to design ncRNAs with specific structures and functions. This could lead to the creation of synthetic ncRNAs optimized for particular tasks, such as precise gene regulation or epigenetic modifications.

3. **Synthetic Biology and Bioinformatics Integration:** The integration of synthetic biology and bioinformatics tools will enable the construction of synthetic gene networks involving ncRNAs. This could lead to the

development of crops with highly engineered and predictable traits (Sang and Chanseok 2016).

Challenges in ncRNA Applications

1. **Off-Target Effects:** Ensuring the specificity of ncRNA-mediated regulation is crucial to avoid unintended off-target effects. Developing robust computational tools and experimental validation methods will be necessary to minimize off-target effects in engineered crops.

2. **Regulatory Concerns:** The release of genetically modified organisms (GMOs), including ncRNA-modified crops, is subject to regulatory approval. Establishing guidelines for the assessment of safety and environmental impacts of ncRNA-based approaches will be important to facilitate their adoption.

3. **Delivery Methods:** Efficient and targeted delivery of ncRNAs into plant cells remains a challenge. Developing effective delivery methods that ensure the uptake and stability of engineered ncRNAs will be essential for successful applications.

4. **Ethical Considerations and Public Perception:** As with any biotechnological advancements, ethical considerations and public perception will play a significant role in the adoption of ncRNA-based technologies. Clear communication, transparency, and engagement with stakeholders will be essential to address concerns and promote acceptance.

Conclusion

The future of ncRNA research holds immense promise for revolutionizing crop improvement. By exploring emerging technologies, refining delivery methods, and addressing challenges, researchers can unlock the full potential of ncRNAs to enhance crop traits, increase resilience, and contribute to global food security (Balyan *et al.*, 2020). As these technologies evolve, the integration of ncRNA-based approaches into conventional breeding programs will shape the next generation of agriculture.

REFERENCES

Ahmed, W., Xia, Y., Li, R., Bai, G., Siddique, K. H., and Guo, P. (2020). Non-coding RNAs: Functional roles in the regulation of stress response in Brassica crops. *Genomics, 112*(2), 1419-1424.

Ariel, F., Romero-Barrios, N., Jégu, T., Benhamed, M., and Crespi, M. (2015). Battles and hijacks: noncoding transcription in plants. *Trends in plant science, 20*(6), 362-371.

Balyan S, Joseph SV, Jain R, Mutum RD, Raghuvanshi S (2020) Investigation into the miRNA/52 isomiRNAs function and drought mediated miRNA processing in rice. *Funct Integr Genomics* 20: 509–522.

Diriba Guta Dekeba. "Advances on the application of non-coding RNA in crop improvement" (2021): *African Journal of Biotechnology* 20(11), 440-450.

Garcia-Ruiz, H.; Takeda, A.; Chapman, E.J.; Sullivan, C.M.; Fahlgren, N.; Brempelis, K.J.; Carrington, J.C. Arabidopsis RNA-dependent RNA polymerases and dicer-like proteins in antiviral defense and small interfering RNA biogenesis during Turnip Mosaic Virus infection. *Plant Cell* 2010, 22, 481–496.

Gull, A., Lone, A. A., and Wani, N. U. I. (2019). Biotic and abiotic stresses in plants. *Abiotic and biotic stress in plants*, 1-19.

Hangauer MJ, Vaughn IW, McManus MT (2013). Pervasive transcription of the human genome produces thousands of previously unidentified long intergenic noncoding RNAs. *PLoS genetics* 9(6): e1003569

He J, Jiang Z, Gao L, You C, Ma X, Wang X, Xu X, Mo B, Chen X, Liu L (2019) Genome-wide transcript and small RNA profling reveals transcriptomic responses to heat stress. Plant Physiol 81: 609–629.

Hua Y, Zhang C, Shi W, Chen H (2019) High-throughput sequencing reveals microRNAs and their targets in response to drought stress in wheat (*Triticum aestivum* L.). *Biotechnol Biotechnol Equip* 33: 465–471.

Hung, T.; Chang, H.Y. Long noncoding RNA in genome regulation: Prospects and mechanisms. *RNA Biol.* 2010, 7, 582–585.

Jiang, P.; Lian, B.; Liu, C.; Fu, Z.; Shen, Y.; Cheng, Z.; Qi, Y. 21-nt phasiRNAs direct target mRNA cleavage in rice male germ cells. *Nat. Commun.* 2020, 11, 5191.

Khraiwesh B, Zhu JK, Zhu J (2012) Role of miRNAs and siRNAs in biotic and abiotic stress responses of plants. Biochim Biophys Acta 1819: 137–148.

Leonetti, P.; Miesen, P.; van Rij, R.P.; Pantaleo, V. Viral and subviral derived small RNAs as pathogenic determinants in plants and insects. *Adv. Virus Res.* 2020, 107, 1–36.

Mercer, T.R.; Dinger, M.E.; Mattick, J.S. Long non-coding RNAs: Insights into functions. *Nat. Rev. Genet.* 2009, 10, 155–159.

Mittler R. Abiotic stress, the field environment and stress combination. *Trends Plant Sci.* 2006;11: 15–19.

Santosh, Baby, Akhil Varshney, and Pramod Kumar Yadava. "Non coding RNAs: biological functions and applications." *Cell biochemistry and function* 33, no. 1 (2015): 14-22.

Sang YS, Chanseok S (2016). Regulatory non-coding RNAs in plants: potential gene resources for the improvement of agricultural traits. *Plant Biotechnology Reports* 10(2): 35-47.

Shin, Sang-Yoon, and Chanseok Shin. "Regulatory non-coding RNAs in plants: potential gene resources for the improvement of agricultural traits." *Plant Biotechnology Reports* 10 (2016): 35-47.

Sailaja Bhogireddy, Satendra K. Mangrauthia, Rakesh Kumar, Arun K. Pandey, Sadhana Singh, Ankit Jain, Hikmet Budak, Rajeev K. Varshney, and Himabindu Kudapa. "Regulatory non-coding RNAs: a new frontier in regulation of plant biology." *Functional and Integrative Genomics* 21 (2021): 313-330.

Peedicayil, J. (2021). Non-coding RNAs and psychiatric disorders. In *Epigenetics in Psychiatry* (pp. 321-333). Academic Press.

Ponting, C.P.; Oliver, P.L.; Reik, W. Evolution and functions of long noncoding RNAs. *Cell* 2009, *136*, 629–641.

Varotto, S., Tani, E., Abraham, E., Krugman, T., Kapazoglou, A., Melzer, R.,. and Miladinoviæ, D. (2020). Epigenetics: possible applications in climate-smart crop breeding. *Journal of Experimental Botany, 71*(17), 5223-5236.

Wang, D.; Qu, Z.; Yang, L.; Zhang, Q.; Liu, Z.H.; Do, T.; Adelson, D.L.; Wang, Z.Y.; Searle, I.; Zhu, J.K. Transposable elements (TEs) contribute to stress-related long intergenic noncoding RNAs in plants. *Plant J.* 2017, *90*, 133–146

Wang, K.C.; Chang, H.Y. Molecular mechanisms of long noncoding RNAs. *Mol. Cell* 2011, *43*, 904–914.

Zhang, P., Wu, W., Chen, Q., and Chen, M. (2019). Non-coding RNAs and their integrated networks. *Journal of integrative bioinformatics, 16*(3), 20190027.

Zhang, X., Du, M., Yang, Z., Wang, Z., and Lim, K. J. (2023). Biogenesis, Mode of Action and the Interactions of Plant Non-Coding RNAs. *International Journal of Molecular Sciences, 24*(13), 10664.

Zhu QH, Wang MB (2012) Molecular functions of long non-coding RNAs in plants. Genes 3: 176–190

Chapter 10

Identifying Novel Mutants: Allele Mining for Harnessing Novel Traits

Nitesh Kushwaha[1], Kanchan Kumari Gupta[2], Digvijay Singh[3] and Amrita Thomas[1]*

[1]*Ph.D, Division of Genetics, ICAR-IARI, New Delhi*
[2]*Ph.D, Division of Vegetable Science, Assam Agricultural University, Jorhat, Assam*
[3]*Assistant Proffesor, Department of Genetics and Plant Breeding, Narayan Institute of Agricultural Sciences, Gopal Narayan Singh University, Jamuhar, Sasaram*
**e-mail: farmernk.988@gmail.com*

INTRODUCTION

Plant breeding is an ever-evolving process which aims at developing new plant types and variety for survival of mankind. The distinctness in the newly developed variety is due to accumulation various alleles existing in germplasm used in the development of variety. Since ages, from the beginning of agriculture man has tried to alter the genetic make up of plant to enhance its yield and agronomic attributes. Unknowingly they incorporated or exploited different alleles or genes. But since the science of genetics began there has been tremendous increase in the manipulation of plant genetic architecture as it unravels the science behind it. But the germplasm used were limited so alleles used were also less. Present day crops have been derived from their ancestral wild form, simultaneously natural gene introgression and natural selection has led to evolution of several landraces and wild relatives which are great sources of novel alleles, these under-utilized alleles can be used to develop agronomically superior (McCouch *et al.*, 2007) and climate smart crops seeking to current situation of climate change and population explosion.

We already know that gene is part of DNA which is involved in development of phenotype in general. And alternative form of gene is called allele. These alleles are nothing mutants of wild from form of gene and result in distinct phenotype from wild form. Mutation is heritable change in genetic makeup of organism which occur spontaneously at lower in frequency in nature. However, most of the mutations are deleterious but still some are beneficial. These beneficial mutations cause development of new gene/allele, ultimately leading in creation of vast germplasm resources in a crop species. Hence, these enormous germplasm resources need to be resorted for novel alleles to further uplift the genetic capability of crop varieties for various agronomic traits.

Recent advances in field of genomics and DNA sequencing in last 15 years has led deposition of enormous amount of sequencing data in various crop database (Chan, 2005; Mardis, 2008). Rapid discovery and annotation of genes and allele can be expected from these sequence information and expression which can be used to design allele-specific markers (Spooner *et al.*, 2005).Based on gene and genome sequences, polymerase chain reaction (PCR) strategies are devised to isolate useful alleles of genes from a wide range of species (Latha *et al.*, 2004).This capability enables direct access to important alleles conferring special characteristic like biotic and abiotic stresses, better nutrient use efficiency, high yield, and improved quality. So using genomic tools researcher can dissect the allelic variants of gene governing a particular trait in different genotypes, this is often referred as 'allele mining'.

In this chapter we will discus the concept of allele mining and its approaches along with its potential and challenges.

Allele and Allele Mining

Alleles are defined as an alternative form of gene and are located on same locus on a homologus chromosome. They are responsible for expression of different forms of a given trait. Allele mining is seeking novel superior alleles for trait in a natural population through latest advances in genomics.

'True' Allele Mining

Earlier studies on allele mining were primarily focused on identification of SNP/InDels at the coding sequence or exonic region of the gene as they are expected to influence the encoded protein structure or phenotype. But recent studies suggests that nucleotide changes in non-coding regions (52 UTR) including promoter, introns and (32 UTR)also have significant influence on transcript synthesis and accumulation which alters the trait expression. Function of intronic mutations in regulation of gene has been evident in expression of genes like tubulin gene and (Fiume *et al.*, 2004) and *rubi3* (polyubiquitin gene) (Samadder *et al.*, 2008) in rice and in wheat and barley in vernalization gene

VRN-1(Fu *et al.*, 2005). In waxy (Wx) gene of rice tenfold increase in gene activity was repoterted which was due to mutation in 52 splice site of the first intron of the gene ((Isshiki *et al.*, 1998). in present scenario, mining the polymorphism in sequences of regulatory region of gene is getting more importance in view gene expression. Now it is well reported that promoter element plays key role in regulation of gene and any changes in promoterregion will alter the gene expression. Now this approach of mining the regulatory elements of genes is called 'promoter mining'.

Development of Allele Mining Set

The germplasms present in the gene bank are evaluated for its morphological characters based on descriptors as well as molecular characterization is done with help of molecular marker. Diversity analysis is carried out using the morphological characterization data and molecular marker data. The germplasm is grouped into clusters depending upon similarity. Genotypes which share common agroclimatic zone will mostly cluster together. Create a core collection by selecting 10 percent of accessions from each cluster or one plant out of 10 if accessions are less. Then collate the core collection with whole germplasm collection for mean, variance, genetic diversity, etc. If these stats between core collection and entire germplasm does not differ significantly, it is versed that the genetic diversity present in core collection is correct representative of entire germplasm. Then the genotypes of core collection are grown in replication at several locations and are evaluated for agronomic, morphological, and qualitative traits to identify better parents. Further these genotypes are sub-grouped and mini-core collection is created. This mini core collection could be used for allele mining.

Approaches of Allele Mining

There are two major approaches for allele mining for identification of sequence variation in given gene in a natural population *viz.* EcoTilling and sequencing-based allele mining.

EcoTilling

TILLING is a method that can recognise polymorphism resulting from induced mutation in candidate gene by heteroduplex analysis (Till *et al.*, 2003). Modified form of TILLING, also called EcoTilling, is means to elucidate extent of natural variation in target genes in crops. Method is nearly same as that of TILLING but in EcoTilling we utilize natural variation occurring in primary and secondary gene pools unlike TILLING which uses induced mutation (Comai *et al.*, 2004; Comai and Henikoff, 2006). EcoTilling also depends upon enzymatic cleavage of heteroduplex DNA with a single stranded specific nuclease (*i.e.*, mung bean nuclease, S1 nuclease, etc.) under certain specific condition then proceeding for variation detection through Li-Cor genotypes (Li-Cor, USA). Nuclease will cleave the strand at site of point mutation. The presence of any variation or point mutation or SNP will be confirmed by sequencing the amplicon from test genotype that carry mutation. Although these techniques were proposed to be cost effective but they are highly sophisticated techniques, require many steps and condition for smooth activity for nuclease activity.

Sequencing Based Allele Mining

This technique entails amplification of alleles in varied genotypes *via.* PCR followed by identification of variation in nucleotide. It helps to analyze individuals for diversity and haplotypes to infer genetic association studies. This approach is much easier and convenient than TILLING and EcoTilling but is cost intensive. But due advancement in sequencing *i.e.* next generation sequencing the procedure has become much more easier and cheaper. 'Massively parallel' methods have emerged with high throughput and accuracy. These methods are mainly required for resequencing, sequence data alignment and reference genome comparison. 454 life sciences first commercialized this kind of method of next generation sequencing (Margulies *et al.*, 2005) and it depends upon technique of pyro-sequencing, eliminating the need of cloning. This sequencing method produced 100 Mb of sequence with accuracy 99.5 per cent accuracy and has average read length of over 250 bases. Another massively parallel sequence known as Illumina/Solexa has been developed (Bennett, 2004) and this has capacity of sequencing one billion bases (1Gb) of 30-40 bases sequence reads in a single run, in a small time of 3-4 days. Applied biosystems Inc (USA) has also developed supported oligonucleotide ligation and detection system (SOLiD) (Hutchison, 2007) another massively parallel sequencer. These advance methods have reduced the cost of sequencing with better accuracy. Moreover, sequencing based allele mining of a particular gene in known accessions and their association with phenotypic variations will give a tremendous impulsion to precision breeding programs in crop plants.

Different Steps Involved in Two Approaches of Allele Mining

Selection of target trait

Identification of accessions associated with desired phenotypic trait

Selection of genes underlying the chosen target trait

Primer designing for whole length of gene

PCR amplification from identified accessions

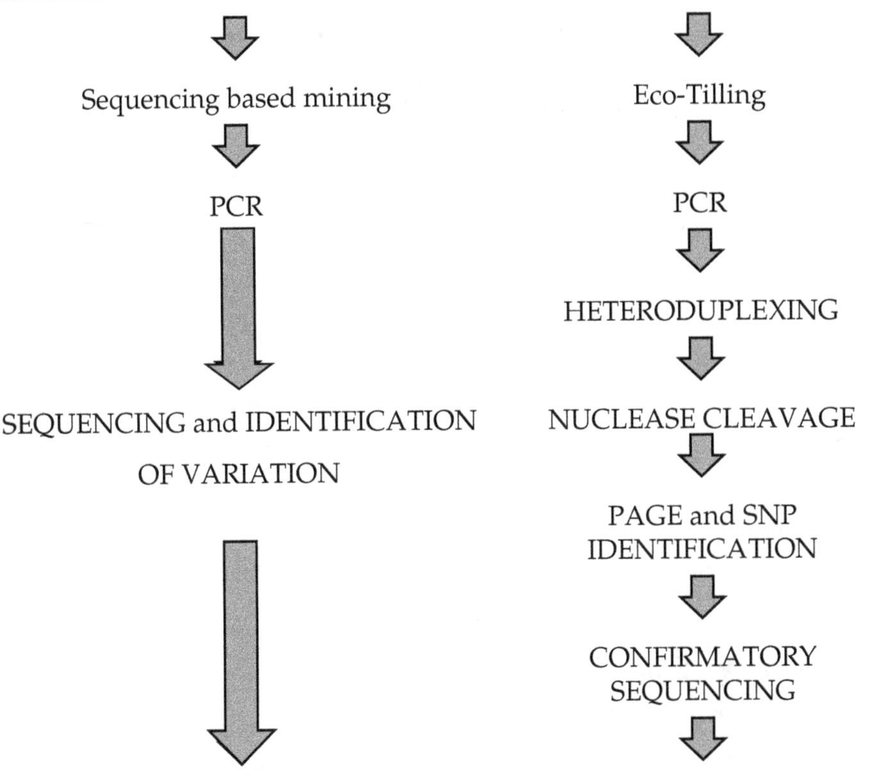

Table 10.1: Comparison between EcoTILLING and Sequence Based Allele Mining

Sl.No.	Particulars	EcoTilling	Sequence Based Allele Mining
1	Technical expertise	Requires high technical expertise.	Require less technical expertise
2	Complexity	High	low
3	Efficiency	Less efficient due to formation of false positive, non-specific cleavage.	Higly efficient as it can identify sequence in one step
4	Utility	More effective in identification of SNPs as compared to InDels	Can detect both SNPs and InDels effectively
5	Cost	Comparatively high	Comparatively low
6	Time	Requires more time	Comparatively less time required
7	Throughput	Less throughput	High throughput

Bioinformatic Tools Required for Allele Mining

Allele mining requires various sophisticated bioinformatic tools *viz.*, PLACE, plantCARE, TRANSFAC, JASPAR, MEME, Plantprom DB, DCPD, SCPD, BioEdit, ClustalW etc. These tools useful for sequence alignment in order to compare new genome sequence to reference genome (Figure 10.1).

Consideration for Allele Mining

Efficient and successful allele mining activity relies mainly on

- ☆ Existing genetic base
- ☆ Availability of genome and gene sequences of a particular crop species
- ☆ Efficient phenotyping technique
- ☆ Genomic resources
- ☆ Cheaper sequencing platform
- ☆ Efficient bioinformatic tools for detection of nucleotide variation.
- ☆ With these criteria allele mining can be initiated for any crop with sequence information.

Application of Allele Mining

There are numerous applications of allele mining *viz.*

- ☆ Characterization of allelic diversity in gene banks
- ☆ Identification of new haplotypes
- ☆ Development of allele-specific markers for MAS
- ☆ Allelic synteny and evolutionary relationship.

Characterization of Allelic Diversity

Identification and retrieval of variation in alleles that influence the phenotype of plant is prime importance for utilization germplasms in crop improvement. Usage of gene bank for effective utilization depends of knowledge of genetic diversity and allelic diversity of candidate gene(s) of interest. Therefore, characterization of allelic diversity and genetic diversity within the existing germplasm is of very much importance to determine the genetic relationships among them and estimate their genetic worthiness in relation to their utility in improving a target trait. Allele mining promises its utility to unlock the genetic diversity in the gene banks (Kaur *et al.*, 2008a). The availability of the sequence of the gene in study that governs a trait will allow practical gene -based use of germplasm (Graner, 2006). Apprehending the capacity of allele mining in PGR management, many national and international institute are now utilizing the facility of allele mining in deciphering the diversity.

Name	Utility	Available at	Reference	Name	Utility	Available at	Reference
PLACE (plant cis acting regulatory DNA elements)	Database of Transcription Factor (TF) binding motifs	http://www.dna. affrc.go.jp/PLACE/ index.html	Higo et al. (1999)	SCPD	Saccharomyces cerevisiade promoter database	http://rulai.cshl. edu/SCPD/	Zhu and Zhang (1999)
plant CARE	Plant cis acting regulatory elements database	http// bioinformatics. psb.urgent.be/ webtools/ plantcare/html	Lescot et al	DCPD	Drosophila core promoter database	http://www- biology.ucsd.edu/ labs/Kadonaga/ DCPD.html	Kutach and Kadonaga (2000)
TRANSFAC	Database of TF and TF binding motifs	http://www. gene-regulation. com/pub/ programs.html	Matyse et al. (2003)	TRED (transcriptional regulatory element (database)	Collection of mammalian regulatory elements	http://rulai.cshl. edu/cgi-bin/ TRED/tred.cgi? process=home	Jiang et al. (2007)
JASPAR	Transcription Factor Binding Site Database	http://jaspar. genereg.net/	Bryne et al. (2008)	ooTFD	Object oriented -transcription factors database	http://www.ifti. org/ootfd/	Ghose (1998)
W-AlighACE	Motif discovery tool	http://www1. spms.ntu.edu.sg/ ~chenxin/W- AlignACE/	Chen et al. (2008a)	AGRIS	TF and RE database	http:// arabidopsis.med ohio-state.edu	Davuluri et al. (2003)
MEME (Multiple EM Elicitation)	Motif discovery tool	http://meme, nbcr.net/ bin/meme.cgi	Bailey et al. (2006)	ModuleFinder and CoReg	Gene co-expression and link with promoter elements	–	Holt et al. (2006)
Plantpron DB	Plant promoter database	http://mentel.cs rhul.ac.uk/ mendel.php? topic=platprom	Shahmuradov et al. (2003)	FastPCR	Nucleotide sequence analysis and primer design	–	Kalendar (2009)
<AToms[ectpr	Tp [redoct TFBS and of promoter analysis	http://www. genomatix.de/ products/ MatInspector/	Cartharius et al. (2005)	Primer 3	Primer design	http://frodo.wi mit.edu/primer3/	Rozen and Skaletsky (2000)
EPD	Eukaryotic Promoter Database	http://www.epd. isb-sib.ch	Schmid et al. (2004)	BioEdit	Nucleotide sequence analysis	www.mbio.ncsu. edu/BioEdit/ BioEdit.html	–
TRRD	Regulatory regions database (description of Regulatory Elements and TFBS)	http://wwwmgs. bionet.nsc.ru/ mgs/gnw/trrd/	Kolchanov et al. (2000)	Clustral W	Sequence alignment	www.ebi.ac.uk/	–

Source: Ashkani *et al.*, 2015.

Figure 10.1: List of some Bioinformatic Tools Used in Allele Mining.

Figure 10.2: Applications of Allele Mining.

Identification of New Haplotypes

Allele mining can be used to determining the extent occurrence of new haplotype, as well as its frequency. The identification of variation in nucleotide sequence in a gene can be done with help of allele mining. Knowledge of frequently occurring haplotype changes and their frequency in the populations can serve as basis of association mapping studies.

Development of Allele-Specific Markers for MAS

Deciphering the variation in sequence will provide way to design and development of allele-specific marker for precise introgression of the identified superior and novel allele in a suitable genetic background. Recently, several case studies have revealed existence of sequence variation at key genes and their utilization of these variations in development of specific markers of MAS. For example, studies of sequence variation in Waxy gene in 18 different accession revealed occurrence of 5 different alleles, which affected the phenotype (Mikami *et al.*, 2008). Allele mining at Pi-ta locus (responsible for resistance to rice blast disease) identified 18 InDel and 91 nucleotide polymorphisms variations among wild rice accessions of *Oryza barthii* and *Oryza rufipogon* (Huang *et al.*, 2008). Successful cases of allele mining to find novel alleles and allelic combinations are also increasingly accumulating in many crops.

Allelic Synteny and Evolutionary Relationship

Allele mining has been used to deduce the sequence information in many crop species, this sequence information can be used to infer synteny between different species and genera. In rye, superior homologue alleles for tolerance to aluminium were identified using syntenic allele sequence information from wheat, and the same technique has been employed to isolate agronomically superior alleles in *Phaseolus vulgaris* and other grasses (Fontecha *et al.*, 2007). Genes governing blast resistance to rice and barley showed remarkable similarity in their sequence and physical position showing common evolutionary origin for blast resistance (Chen *et al.*, 2002).

Some Examples of Allele Mining in Crop Improvement

Allele Mining for Blast Resistance Genes in Finger Millet and Rice

Finger millet is a rich source of nutrient but its one major constraint is blast disease caused by *Magnaporthe grisea,* which also is pathogen of rice. It causes severe yield losses in both the crops (Srinivasachary *et al.*, 2007). Therefore, there is need to know the molecular mechanism of disease resistance and identify the underlying gene responsible for resistance. As finger millet sequencing has not been done yet on the other hand sequence information of rice well known. Comparative genomics is of utmost importance to dissect agronomically important traits in finger millets. Study of Srinivasachary *et al.* (2007) performed comparative analysis of rice and finger millet and reported significant synteny between them.

Allele Mining for Drought Tolerance in Barley

Haplotype and sequence variation(SNP/INDELs) were identified at 9 loci in 96 barley accessions namely (*HvARH1, HvSRG6, HvDRF1, HVA1, HvDREB1, HvNHX1, HVP1, HvNud and HvPPRPX*) which were involved in stress response pathways. The analysis consisted sequence information of about 1.5 million base pairs in barley and the identified haplotypes and polymorphisms were recorded in a database named BAHADAS ("BArleyHAplotypeDAtabaSe for drought-related candidate genes").

Eco-Tilling for Salinity Tolerance Genes in Barley Cultivars

Adnan Al-Yassin and Raheleh Khademian (2015) used EcoTILLING approach in barley fordiscovery and detection of DNA polymorphisms for salt stress tolerance. It was concluded that it is possible to combine different alleles from wild barley to obtain cultivars with more salt stress tolerance.

Table 10.2: Status of Allele in some Crop Plants

Crop	Allels/Locus	Trait/Protein
Wheat	*Viviparous-1*	Pre-harvest sprouting tolerance
rye	ALT3	Aluminium tolerance
Apple	*Mal d 3*	Allergenicity
Apricot	**S**	Self-incompatibility
Barley	*Amy32b*	α amylase
Barley and wheat	*transcription factor -GAMYB*	GAMYB-involved in gibberellin signaling
Barley	*VRN-H1 and VRN-H2*	Vernalization requirement
Barley	*Bmy1*	β-amylase I- starch break down enzyme
Tomato	**Pto**	Disease resistance
Barley	*Gpc-B1*	Grain protein content
Barley	*rps2*	Ribosomal protein S2
Potato	*Rpi-blb1*	Late blight resistance
Rice	*Badh2*	Fragrance
Rice	*Pi ta*	Blast resistance

Major Constraint in Allele Mining

Selection of Genotypes

The most difficult challenge to unlock the existing variation is germplasm selection to be mined. It is very difficult to manage large number of germplasm and to genotype them with existing facilities. There is a need to find a way for screening the germplasm efficiently to discover new allele and their combinations within few accessions rather than large number of genotypes. Development of core collections, precise and efficient phenotyping methods and flexible computational tool can be used to overcome this challenge.

Flexible Computational Tools

Rapid increase in number of plant genetic resources in gene bank has led to problem of genetic verbosity and repetitiveness within a gene pool of a crop species, which has created a problem in managing the resources as well their utilization. So, in order to mitigate this challenge robust computational tools are needed to gain access to useful alleles which itself is one of the major challenges in allele mining. A computational tool called 'Power Core' has been developed by Park (2007) to guide the development of core and allele mining sets by improving the richness and decreasing the repetitiveness of useful alleles. Tools like this need to be developed to accelerate the process of allele mining.

Handling Genomic Resources

To match the speed of accumulation of gene expression and nucleotide data, statistical computational tools need to be created to decipher the functional nucleotide diversity and to foresee nucleotide changes responsible for altered function. Utilizing the development in allele mining, association studies and comparative genomics by combining expertise from multiple discipline like bioinformatics, molecular biology and statistics is a way to address this problem.

Demarcation of Promoter Region

Promoter and regulatory sequences are located upstream of gene coding sequence and position varies from gene to gene. Existing literature clearly reflects the absence of consensus in determining the size of the region for mining promoter and regulatory elements. This size varies from –100 bp to –3000 bp (Brazma *et al.*, 1998; Molina and Grotewold, 2005; Park *et al.*, 2002; Veerla and Hoglund, 2006). The promoter mining is a tedious task and specific software need to be developed to know the exact position and sequences of the core promoter region.

Characterization of Regulatory Elements

Characterization of coding sequence is quite easy but on the other hand characterization of extent of variation in cis-acting regulatory region poses a great challenge. It is difficult to identify it even in fully sequenced genome. Trans-acting factors or environmental fluctuation limits the screening of regulatory variants based of transcript level differences (Dai *et al.*, 2007). Prediction of orientation of promoter will be difficult even by software as they can work in bidirectional pathway. Analyzing these promoters becomes more difficult as transcriptional factors usually control more than one gene (Hehl and Wingender, 2001). To overcome this problem, it is suggested to utilize services of different computational methods to get the picture of regulatory sequences which can be further analyzed through suitable experiment.

Higher Sequencing Costs

Cheaper and effective sequencing method need to be developed to decrease the cost of whole genome sequencing. Also, labour and time saving aspect of sequencing need to be addressed to make over all allele mining as a successful approach to identify novel and mutant allele which can be used in crop improvement program.

Conclusion

Allele mining is new approach to rapidly identify new gene and their mutant allele which helps to support crop improvement programs effectively. Allele mining provide insight into molecular basis of traits and gives information about nucleotide sequence changes, associated with occurrence of superior allele. Allele mining has various application like elucidating genetic diversity, tracing evolutionary history and synteny, identification of new haplotype and development of allele specific marker in MAS. Seeking the potentiality of allele mining many crop research institutes at international level have used to discover new alleles in various crop species like wheat, rice, barley, pea, soyabean, tomato, etc. Though it has many applications, it has certain limitation which have been discussed earlier in this chapter and those need to be addressed to make it a successful approach for crop improvement.

REFERENCES

Al-Yassin, A., and Khademian, R. (2015). Allelic variation of salinity tolerance genes in barley ecotypes (natural populations) using EcoTILLING: a review article. *Am Eur J Agric Environ Sci*, *15*(4), 563-572.

Bailey, T. L., Williams, N., Misleh, C., and Li, W. W. (2006). MEME: discovering and analyzing DNA and protein sequence motifs. *Nucleic acids research*, *34*(suppl_2), W369-W373.

Bennett, S. (2004). Solexa ltd. *Pharmacogenomics*, *5*(4), 433-438.

BJ, T. (2003). Large-scale discovery of induced point mutations with high-throughput TILLING. *Genome Res*, *13*, 524-530.

Brâzma, A., Jonassen, I., Vilo, J., and Ukkonen, E. (1998). Predicting gene regulatory elements in silico on a genomic scale. *Genome research*, *8*(11), 1202-1215.

Bryne, J. C., Valen, E., Tang, M. H. E., Marstrand, T., Winther, O., da Piedade, I.,. and Sandelin, A. (2007). JASPAR, the open access database of transcription factor-binding profiles: new content and tools in the 2008 update. *Nucleic acids research*, *36*(suppl_1), D102-D106.

Cartharius, K., Frech, K., Grote, K., Klocke, B., Haltmeier, M., Klingenhoff, A.,. and Werner, T. (2005). MatInspector and beyond: promoter analysis based on transcription factor binding sites. *Bioinformatics*, *21*(13), 2933-2942.

Chan, E. Y. (2005). Advances in sequencing technology. *Mutation Research/ Fundamental and Molecular Mechanisms of Mutagenesis*, *573*(1-2), 13-40.

Chen, H., Wang, S., Xing, Y., Xu, C., Hayes, P. M., and Zhang, Q. (2003). Comparative analyses of genomic locations and race specificities of loci for quantitative resistance to Pyricularia grisea in rice and barley. *Proceedings of the National Academy of Sciences*, *100*(5), 2544-2549.

Chen, X., Guo, L., Fan, Z., and Jiang, T. (2008). W-AlignACE: an improved Gibbs sampling algorithm based on more accurate position weight matrices learned from sequence and gene expression/ChIP-chip data. *Bioinformatics*, *24*(9), 1121-1128.

Comai, L., and Henikoff, S. (2006). TILLING: practical single nucleotide mutation discovery. *The Plant Journal*, 45(4), 684-694.

Comai, L., Young, K., Till, B. J., Reynolds, S. H., Greene, E. A., Codomo, C. A.,. and Henikoff, S. (2004). Efficient discovery of DNA polymorphisms in natural populations by Ecotilling. *The Plant Journal*, 37(5), 778-786.

Dai, L. Y., Liu, X. L., Xiao, Y. H., and Wang, G. L. (2007). Recent advances in cloning and characterization of disease resistance genes in rice. *Journal of Integrative Plant Biology*, 49(1), 112-119.

Davuluri, R. V., Sun, H., Palaniswamy, S. K., Matthews, N., Molina, C., Kurtz, M., and Grotewold, E. (2003). AGRIS: Arabidopsis gene regulatory information server, an information resource of Arabidopsis cis-regulatory elements and transcription factors. *BMC bioinformatics*, 4, 1-11.

de Vicente, M. C., and Glaszmann, J. C. (2006). *Molecular markers for allele mining*. Bioversity International.

Fiume, E., Christou, P., Gianì, S., and Breviario, D. (2004). Introns are key regulatory elements of rice tubulin expression. *Planta*, 218, 693-703.

Fontecha, G., Silva-Navas, J., Benito, C., Mestres, M. A., Espino, F. J., Hernández-Riquer, M. V., and Gallego, F. J. (2007). Candidate gene identification of an aluminum-activated organic acid transporter gene at the Alt4 locus for aluminum tolerance in rye (Secale cereale L.). *Theoretical and Applied Genetics*, 114, 249-260.

Fu, D., Szûcs, P., Yan, L., Helguera, M., Skinner, J. S., Von Zitzewitz, J.,. and Dubcovsky, J. (2005). Large deletions within the first intron in VRN-1 are associated with spring growth habit in barley and wheat. *Molecular genetics and genomics*, 273, 54-65.

Gokidi, Y., Bhanu, A. N., Chandra, K., Singh, M. N., and Hemantaranjan, A. (2017). Allele mining–an approach to discover allelic variation in crops. *The Journal of Plant Science Research*, 33(2), 167-180.

Hehl, R., and Wingender, E. (2001). Database-assisted promoter analysis. *Trends in Plant Science*, 6(6), 251-255.

Higo, K., Ugawa, Y., Iwamoto, M., and Korenaga, T. (1999). Plant cis-acting regulatory DNA elements (PLACE) database: 1999. *Nucleic acids research*, 27(1), 297-300.

Holt, K. E., Harvey Millar, A., and Whelan, J. (2006). ModuleFinder and CoReg: alternative tools for linking gene expression modules with promoter sequences motifs to uncover gene regulation mechanisms in plants. *Plant Methods*, 2(1), 1-15.

Huang, C. L., Hwang, S. Y., Chiang, Y. C., and Lin, T. P. (2008). Molecular evolution of the Pi-ta gene resistant to rice blast in wild rice (Oryza rufipogon). *Genetics*, 179(3), 1527-1538.

Hutchison III, C. A. (2007). DNA sequencing: bench to bedside and beyond. *Nucleic acids research, 35*(18), 6227-6237.

Isshiki, M., Morino, K., Nakajima, M., Okagaki, R. J., Wessler, S. R., Izawa, T., and Shimamoto, K. (1998). A naturally occurring functional allele of the rice waxy locus has a GT to TT mutation at the 52 splice site of the first intron. *The Plant Journal, 15*(1), 133-138.

Kalendar, R., Lee, D., and Schulman, A. H. (2009). FastPCR software for PCR primer and probe design and repeat search. *Genes, Genomes and Genomics, 3*(1), 1-14.

Kaur, N. (2008). *Allele mining and sequence diversity at the wheat powdery mildew resistance locus Pm3* (Doctoral dissertation, University of Zurich).

Kim, K. W., Chung, H. K., Cho, G. T., Ma, K. H., Chandrabalan, D., Gwag, J. G.,. and Park, Y. J. (2007). PowerCore: a program applying the advanced M strategy with a heuristic search for establishing core sets. *Bioinformatics, 23*(16), 2155-2162.

Kutach, A. K., and Kadonaga, J. T. (2000). The downstream promoter element DPE appears to be as widely used as the TATA box in Drosophila core promoters. *Molecular and cellular biology, 20*(13), 4754-4764.

Latha, R., Rubia, L., Bennett, J., and Swaminathan, M. S. (2004). Allele mining for stress tolerance genes in Oryza species and related germplasm. *Molecular biotechnology, 27*, 101-108.

Lescot, M., Déhais, P., Thijs, G., Marchal, K., Moreau, Y., Van de Peer, Y.,. and Rombauts, S. (2002). PlantCARE, a database of plant cis-acting regulatory elements and a portal to tools for in silico analysis of promoter sequences. *Nucleic acids research, 30*(1), 325-327.

Mardis, E. R. (2008). Next-generation DNA sequencing methods. *Annu. Rev. Genomics Hum. Genet., 9*, 387-402.

Margulies, M., Egholm, M., Altman, W. E., Attiya, S., Bader, J. S., Bemben, L. A.,. and Rothberg, J. M. (2005). Genome sequencing in microfabricated high-density picolitre reactors. *Nature, 437*(7057), 376-380.

Matys, V., Fricke, E., Geffers, R., Gößling, E., Haubrock, M., Hehl, R.,. and Wingender, E. (2003). TRANSFAC®: transcriptional regulation, from patterns to profiles. *Nucleic acids research, 31*(1), 374-378.

McCouch, S. R., Sweeney, M., Li, J., Jiang, H., Thomson, M., Septiningsih, E.,. and Ahn, S. N. (2007). Through the genetic bottleneck: O. rufipogon as a source of trait-enhancing alleles for O. sativa. *Euphytica, 154*, 317-339.

Mikami, I., Uwatoko, N., Ikeda, Y., Yamaguchi, J., Hirano, H. Y., Suzuki, Y., and Sano, Y. (2008). Allelic diversification at the wx locus in landraces of Asian rice. *Theoretical and Applied Genetics, 116*, 979-989.

Park, P. J., Butte, A. J., and Kohane, I. S. (2002). Comparing expression profiles of genes with similar promoter regions. *Bioinformatics, 18*(12), 1576-1584.

Rozen, S., and Skaletsky, H. (1999). Primer3 on the WWW for general users and for biologist programmers. *Bioinformatics methods and protocols*, 365-386.

Samadder, P., Sivamani, E., Lu, J., Li, X., and Qu, R. (2008). Transcriptional and post-transcriptional enhancement of gene expression by the 52 UTR intron of rice rubi3 gene in transgenic rice cells. *Molecular Genetics and Genomics, 279*, 429-439.

Schmid, C. D., Praz, V., Delorenzi, M., Périer, R., and Bucher, P. (2004). The Eukaryotic Promoter Database EPD: the impact of in silico primer extension. *Nucleic acids research, 32*(suppl_1), D82-D85.

Shahmuradov, I. A., Gammerman, A. J., Hancock, J. M., Bramley, P. M., and Solovyev, V. V. (2003). PlantProm: a database of plant promoter sequences. *Nucleic acids research, 31*(1), 114-117.

Srinivasachary, Dida, M. M., Gale, M. D., and Devos, K. M. (2007). Comparative analyses reveal high levels of conserved colinearity between the finger millet and rice genomes. *Theoretical and Applied Genetics, 115*, 489-499.

Veerla, S., and Höglund, M. (2006). Analysis of promoter regions of co-expressed genes identified by microarray analysis. *BMC bioinformatics, 7*, 1-15.

Zhu, J., and Zhang, M. Q. (1999). SCPD: a promoter database of the yeast Saccharomyces cerevisiae. *Bioinformatics (Oxford, England), 15*(7), 607-611.

Chapter 11

Variants in Bulked Segregant Analysis and MutMap

Puja Mandal[1], Meghna Mandal[2], Sharada H.B.[3] and Raiza Christina G.[1]*

[1]*Ph.D. Scholar, Department of Genetics and Plant Breeding, Tamil Nadu Agricultural University, Coimbatore – 641 003, Tamil Nadu*
[2]*Ph.D. Scholar, Department of Plant Breeding and Genetics, Punjab Agricultural University, Ludhiana – 141 004, Punjab*
[3]*Ph.D. Scholar, Department of Genetics and Plant Breeding, Keladi Shivappa Nayaka University of Agricultural and Horticultural Sciences, Shivamogga – 577 201, Karnataka*
**e-mail: pujaman@gmail.com*

ABSTRACT

Mapping genes is fundamental to enable us to unravel the complexities of genetics, biology, agriculture and medicine. It aids in understanding how genes influence traits, diseases, and evolutionary processes, leading to advancements in various scientific and practical fields. Most of the traits which fall into distinct phenotypic classes are usually governed by one major gene and are qualitative traits. Mapping of genes governing qualitative traits and identification of markers linked to them was initially made possible in the form of Bulked Segregant Analysis (BSA). To accommodate various aspects such as experimental flexibility, genetic diversity, marker type and availability, genome structure, trait complexity as well as cost and resource availability, different variants of the same came into being. MutMap is another such forward genetics tool which makes use of mutant populations and advanced sequencing technologies to map a particular gene responsible for a trait. Different variations in Mutmap have also been generated to overcome the limitations in each type. Researchers may choose the most appropriate method or integrate them based on these factors and their requirements to effectively map and understand the genetic basis of specific traits.

Keywords: Mapping, BSA, MutMap, Genes.

INTRODUCTION

The ultimate basis of genetic variation present in any natural population is either due to spontaneous mutations or activation of natural mutagens or exposure to other chemical mutagens in the form of food, drugs etc. In situations where there is little natural variation, artificial mutagens have been successfully utilized to generate genetic variability. These mutagens affect the observable phenotype by introducing SNPs, InDels, or segmental breaks. As hypo- and hypermorphic alleles that are beneficial for studying gene functions and breeding can be created by induced point mutations, they are significant genetic resources.

Identification of trait polymorphism and dissection of underlying genetic loci controlling this trait is the holy grail of agricultural genetics. Genome mapping determines gene positions along chromosomes using Mendelian principles. Genetic linkage maps show gene/marker order. Large mapping populations result from crosses of contrasting parents. Physical mapping estimates actual base pair distances between genes. Traits are usually influenced by multiple distributed loci across the genome. It is now of common consensus that a trait is controlled by a single gene only very rarely, rather, most traits are indeed regulated by multiple loci distributed across the whole genome. So, traits are rightly called quantitative traits, and the genomic locations governing them are called quantitative trait locus (QTL). To precisely detect the genetic control of a trait and then manipulate it, one has to link all loci with that trait (Bhat and Yu, 2021).

The discovery of DNA sequencing by Sanger *et al.*, 1977 was the harbinger of the genomics era. Genomic data with optimum quality can now be generated with automation and reûnement of existing DNA sequencing technologies with significant reduction in time and expenses. Going hand in hand is the improvement in bioinformatics tools and algorithms which is also contributing equally by keeping pace with the enormous amount of data generated by these technologies.

Bulked Segregant Analysis

QTL mapping followed by fine mapping, physical mapping and then positional cloning of genes are powerful and the most used methods to understand the genetic basis of phenotypic variation of agronomically important traits. However, this package of steps, which is referred to as individual segregant analysis (ISA), has some important drawbacks- they are usually time consuming and low-throughput (Song *et al.*, 2017). An alternative approach, named Bulked Segregant Analysis (BSA) is in vogue now. BSA is an ingenious and rapid method for identifying markers linked to any specific casual gene for a phenotype or genomic region. While ISA classifies segregants according to their marker genotype, BSA pools segregants according to their phenotypes (Majeed *et al.*, 2022). While ISA compares trait values of different classes, BSA compares marker allele frequencies in different classes (Huang *et al.*, 2020). Two pools of individuals are created- one which expresses the said phenotype and the other which does not. Allele frequencies of the genetic markers is quantified in these pools/bulks, which gives an indication of the linkage between markers and the causal gene (Liu *et al.*, 2012). BSA performs better than ISA on numerous grounds- it is more precise, simple, quicker and cheaper than ISA. It also overcomes the low-throughput and time-consuming bottlenecks of ISA.

The BSA technique employs a seemingly straightforward approach: it generates two groups with distinct physical traits from the offspring resulting from a cross between individuals displaying extreme traits. These groups, known as 'bulks,' consist of individuals that share a common trait or genomic region but display randomness in unlinked regions. This genetic distinction ensures dissimilarity between the two bulks concerning the selected trait and the specific genomic region while maintaining a random distribution of alleles in other loci. For example, in the creation of 'susceptible' and 'tolerant' bulks, allele 'A' might be prevalent in susceptible plants, and allele 'a' in tolerant plants, while alleles unrelated to the trait segregate randomly in both bulks. Initially, BSA was conceived to develop genetic markers for target traits in early life stages of test organisms (Giovannoni *et al.*, 1991; Michelmore *et al.*, 1991). BSA can be seen as a form of selective genotyping, focusing solely on the extreme ends (individuals with extreme traits) of a population for genotyping (Darvasi and Soller, 1994; Sun *et al.*, 2010). This approach significantly reduces costs and streamlines the analysis, all while maintaining statistical robustness (Majeed *et al.*, 2022). Furthermore, costs are minimized by pooling all individuals selected from each extreme end for joint analysis. To illustrate, in a population of 500 individuals with 25 selected as extreme cases for each bulk, BSA would only consume 0.4 per cent (=2/500) of the total expenses required for analyzing the entire population (Zou *et al.*, 2016).

Various marker systems have been applied in the context of BSA. The sole requirement is that the markers chosen must offer a quantitative assessment of allelic frequencies (Liu *et al.*, 2012). Both markers reliant on hybridization and those reliant on PCR have found use. In recent times, the effectiveness of BSA has been significantly enhanced with the utilization of sequence-based markers, including technologies like whole genome sequencing and markers based on restriction-site associated DNA (RAD).

Classical BSA

The traditional BSA approach, using genetic markers like microsatellites or RFLPs for genotyping the pooled DNA

Bulked Segregant Analysis is an elegant QTL mapping strategy that permits the simultaneous isolation of genes that affect a specific phenotype (Michelmore *et al.*, 1991). Seminal work on BSA utilized an F_2 population to identify RAPD and RFLP markers linked to a trait of interest. Classical BSA is performed using single genetic markers such as repeat-based markers or PCR-based markers. The throughput of classical BSA can be increased by using array- or chip-based genotyping (DART-SNP array, comparative genomic hybridization array, TILLING array *etc.*). The number of markers analysed using array systems is much more that the conventional DNA markers, thereby reducing costs and increasing effectiveness (Zou *et al.*, 2016). This method is convenient when a) simple SSR makers are not accessible in the target region b) the genetic base of the crop is narrow, making detection of polymorphic genetic markers very challenging, as in pigeon pea or groundnut (Zegeye *et al.*, 2018).

Next-Generation Sequencing (NGS)-based BSA

Utilizing high-throughput sequencing technologies to comprehensively analyze genetic variation in the pooled DNA samples, including SNPs, INDELS, and structural variants.

BSA-Seq is a speedy approach to accomplish BSA using next-generation sequencing (NGS) platforms. BSA, in conjugation with NGS enables the rapid identification and dissection of both qualitative and quantitative traits (Zhang *et al.*, 2021). To distinguish significant SNP-trait associations from any spurious associations, comparatively high sequencing depth and coverage are needed. This increases the efficiency of BSA-Seq. However, high sequence coverage and depth increases the sequencing cost, thereby increasing the overall expenditure of BSA-Seq (Zhang *et al.*, 2021), making BSA-seq prohibitory to organisms with large genomes (Tang *et al.*, 2018). However, BSA-seq necessitates only two sequencing reactions for the two extreme bulks, thus compensating for the added costs due to high sequencing depth and coverage. Thus, instead of constructing a large pool, sequencing must be performed to the deepest affordable level - this is the tradeoff necessary for a good BSA-seq experiment (Majeed *et al.*, 2020).

QTL-Seq

Coupling BSA with quantitative trait locus (QTL) analysis to identify regions of the genome associated with quantitative traits.

Most traits of agronomic importance are controlled by polygenes/multiple genes and hence, are quatitative in nature. A single gene has a feeble effect on the phenotype, but the cumulative effect of all the genes generates the quantitative trait variation. QTL mapping is a key strategy for dissecting the genetic loci underlying quantitative traits and has been amply used during plant breeding. Marker assisted QTL mapping has been amply used in crop improvement programmes, as is illustrated by the incorporation of the Sub-1 gene in the MTU-7029 (Swarna) rice background to generate submergence-tolerant rice (Neeraja *et al.*, 2007). QTL mapping has an important pre-requisite- the necessity of molecular markers that can effectively discriminate parental lines, implying that the two parents are chosen from unrelated genetic backgrounds. However, since the trait is governed by many genes, and the contrasting parents are also different for most of the traits, identifying the gene location becomes complicated. To resolve this, one may use parents which are genetically similar, but this would decrease the likelihood of recovery of a polymorphic marker linked to the trait of interest. Moreover, QTL mapping depends on conventional single marker linkage analysis methods, which is both expensive and time consuming. QTL-seq is an effective counter to all these drawbacks of QTL mapping. The efficiency of QTL-Seq for QTL identification has been amply demonstrated by Takagi *et al.* (2013), who applied QTL-seq to a RIL population and F_2 populations and successfully identified QTLs for important agronomic traits, such as seedling vigor and resistance to rice blast disease. The pipeline for QTL-seq, as proposed by Takagi *et al.* (2013), is described below:

1. Generate a mapping population by hybridizing two varieties which differ for the trait of interest. (RILs and DH are the populations of choice due to their high degree of homozygosity, however, F_2 may also be used, since it can be generated in a much smaller time span. Also, replicated evaluation os possible for RILs and DHs, but not F_2).

2. Measure the progeny of the mapping population for the trait of interest. If the trait is quantitative, the frequency distribution of the measured values for the trait will approach a normal distribution.

3. Sample DNA from the10-20 individuals on each extreme of the normal curve.

4. Whole genome resequencing of the bulked DNA at >6x genome coverage.

The expectation is that the bulked DNA will contain genomes from both parents in a 1:1 ratio for the majority of genomic regions. However, in the genomic region causing the phenotypic difference between the two bulks, there would be unequal representation of the parents.

Since QTL-seq is performed using sequencing data, the relative amount of the genomes derived from the two parents can be evaluated from the proportion of short reads corresponding to each parental genome that can be discriminated by SNPs available between the two. The sequence data is aligned to the reference sequence.

An index called SNP-index, which is the number of short reads harbouring SNPs that are different from the reference sequence (k) divided by the total number of short reads (n) has been devised, whose value can range from 0 to 1. A value of 0 indicates that all the short reads contain genomic fragments from the parent that was used as a reference sequence. A value of 1 indicates that all the short reads represent the genome from the other parent.

A SNP-index value of 0.5 signifies an equal contribution of genetic material from both parents to the combined offspring. SNPs displaying an SNP-index below 0.3 are excluded due to their inability to be differentiated from erroneous SNPs caused by sequencing errors or misalignment. The average SNP-index of all SNPs within a specific genomic interval is calculated and then subjected to an analysis that involves sliding windows. In regions of the genome where phenotypic differences between the two groups of progeny are not attributed, the SNP-index graphs should appear identical in both groups. Conversely, in genomic segments responsible for the phenotypic distinction between the two groups, the SNP-indices of these causative regions would exhibit mirrored patterns in relation to the SNP-index line of 0.5. Such regions are anticipated to hold a higher likelihood of containing Quantitative Trait Loci (QTLs) responsible for the observed phenotypic divergence. It is crucial to compare the two graphs in order to differentiate QTLs from genomic areas displaying segregation distortion due to factors other than artificial directional selection. This distortion leads to the SNP-index deviating from 0.5 in the same direction in both groups. As an alternative approach, the two graphs can be merged to generate a ΔSNP index graph by subtracting the SNP index of one group from the other. The range of ΔSNP-index values can extend from -1 to +1, indicating similarity to either parent, while a value of 0 signifies that both parents share the same SNP-indices within the genomic regions.

QTL-seq was effectively used by Takagi *et al.* (2013) to map QTLs governing resistance/susceptibility to rice blast in Japonica cultivars (Hitembore, Nortai). BSA was done on a RIL population and phenotyping (disease reaction after inoculation with a pathogen) was repeated four times over a period of 4 years to ensure accurate phenotyping and efficient bulking. The QTL was mapped to a chromosomal region of 2.39-4.39 Mb on chromosome 6. To verify the candidate QTLs detected by the QTL-seq method, traditional QTL analysis was again carried out using SNP markers. This traditional analysis delimited the QTL to a 0-4.9 Mb region on chromosome 6, thus reinforcing the results obtained using QTL-seq. To judge the practicality of QTL-seq on an F_2 population, an attempt was made to map QTLs governing seedling from a cross between Hitembore (low vigour) and Dunghan Shali (high vigour). Bulking was done in F_2 itself. QTL-seq procedure indicated 2 genomic locations that explained the difference between the two bulks. Both these genomic locations coincided with previously identified QTLs from conventional QTL analysis in an F_7 population. This result demonstrates that QTL-seq in an F_2 population was as effective as traditional QTL mapping using RILs in F_7 generation.

There is a suggestion that QTL-seq has the potential to be utilized across diverse populations to identify genomic areas that have experienced either artificial or natural selection.

RNA-Seq-based BSA: Applying similar principles as NGS-based BSA, but focusing on analyzing gene expression differences between phenotypically distinct RNA pools.

SNPs can be identified from transcriptomic data, hence, it is possible to use RNA-sequencing technology to efficiently identify SNPs from the bulks (Majeed *et al.*, 2020). Bulked segregant RNA-Seq (BSR-Seq) makes use of RNA-Seq reads to efficiently map genes in populations for which no polymorphic markers have been previously identified (Liu *et al.*, 2012). The phenotypic screening and bulking steps are the same in BSR-seq, the only difference being that the transcribed genome is used as the genotypic information.

Since NGS data exists in a digital format, it is possible to go for de novo SNP discovery and genotype BSA samples by analyzing the same RNA-seq data using an empirical Bayesian approach. Additionally, RNA-seq data provides information about the spatiotemporal aspects of gene expression at no extra cost.

Trick *et al.* (2012) presented a proof of concept for using next-generation sequencing (NGS) for SNP discovery in tetraploid wheat lines and successfully fine-mapping the GPC-B1 gene. The study combined SNP discovery from NGS data with bulked segregant analysis to fine-map genes in polyploid wheat. The results demonstrated the potential of NGS-based SNP discovery and bulked segregant analysis for gene mapping in polyploid crops, which could lead to improved breeding and crop yields in the future.

Wang *et al.* (2017) used bulked segregant RNA-Seq (BSR-Seq) to develop SNP markers and map the YrZH22 gene to a genetic interval on wheat chromosome 4BL. specifically, the researchers used BSR-Seq to identify differentially expressed genes between two bulked RNA samples, one from resistant and one from susceptible $F_{5:6}$ RILs. They then used SNP mapping to identify the genomic location of YrZH22 between two markers on chromosome arm 4BL.

In maize, the glossy13 gene was identified using Seq-Walking technology in combination with the mapping interval defined by the BSR-seq experiment. The Seq-Walking was conducted on DNA extracted from pools of homozygous mutant and wild-type relatives. The focus was mainly on the "mutant-specific" MFS (Li *et al.*, 2013).

Significant strides have been made in sugarcane breeding for smut resistance using BSR-Seq. An F_2 population was raised from a cross of smut-resistant variety YT93-159 and a smut-susceptible variety ROC22. Bulks were prepared from the F_2 progeny, and total RNA of individuals in each bulk were pooled and subjected to BSR-seq. DEGs identified were enriched for stress-related metabolic pathways, including plant hormone signal transduction, plant-pathogen interactions, carbon metabolism, phenylalanine metabolism and glutathione metabolism. Candidate genes were identified which were associated with smut resistance as well as stress resistance (Wu *et al.*, 2022).

BSR-seq may not be very effective for traits that are highly affected by environment and the traits governed by many minor genes (Edae and Rouse, 2019).

Pool-Seq

A variant of BSA where genomic DNA from individuals is pooled before sequencing, reducing the cost of sequencing individual genomes.

The decreasing costs of sequencing technologies have significantly benefited all gene mapping endeavors. However, the widespread genomic sequencing of populations is currently limited mainly to model species. As an economical alternative to individual sequencing, the concept of sequencing pooled groups of individuals, known as Pool-seq, has emerged. This approach holds significant promise for both model and non-model species in the field of population genomics.

In surveys of population genetics, two critical parameters come into play: information about polymorphic sites in the genome and allele frequencies across different populations. The accuracy of allele frequency determination significantly impacts the effectiveness of genetic analysis. It's crucial to determine allele frequencies from sizable populations, as smaller samples are more susceptible to higher sampling variations, potentially leading to notable inaccuracies even if the allele frequencies were initially determined reasonably well. Through a straightforward strategy, Pool-seq achieves a noteworthy degree of accuracy in determining allele frequencies while concurrently reducing costs. This is accomplished by utilizing a large population for sampling, but not analyzing every chromosome. Only a minimal number of chromosomes are subjected to multiple rounds of sequencing. Unlike individual sequencing that necessitates a separate library for each genome, Pool-seq employs a single sequencing library for the entire sample. This streamlined approach in library preparation contributes to substantial cost savings due to the considerably reduced number of sequencing libraries required.

Cost efficiency can be attained by utilizing methods like exome sequencing, high-throughput RNA sequencing (RNA-seq), or the application of restriction-site-associated DNA (RAD) markers. However, each of these techniques has its shortcomings. Exome sequencing, for instance, comes with the expense of exome enrichment. Moreover, it displays a bias towards protein-coding segments of the genome. On the other hand, Pool-seq offers a relatively impartial approach, enabling the identification of functional disparities without a specific genomic target. This aspect is crucial to acknowledge, as growing evidence suggests that causal genetic variations tied to significant phenotypes often exist within unexplored genomic regions with regulatory roles. (Karczewski *et al.*, 2013; Khurana *et al.*, 2013; Schaub *et al.*, 2012). Although RNA-seq boasts species independence and is a favoured method, its use in pooled sample DNA polymorphism analysis is hampered by uneven levels of allele-specific gene expression. This results from the impact of cis-regulatory elements, causing variations in relative expression levels among alleles. To address this, Pool-seq could be effectively combined with RNA-seq, especially for species possessing sizable genomes or those currently lacking a reference genome. (Gross *et al.*, 2013; Kozak *et al.*, 2014; Sloan *et al.*, 2012).

Factors determining the power of Pool-seq are:

a) Pool size: since pool-seq is designed to obtain allele frequency estimates from larger samples, applying this method to small pools (<50 individuals) will result in suboptimal output.

b) Linkage disequilibrium: Pool-seq based on short reads data is unsuitable to infer haplotypes and LD. Longer sequencing reads can overcome this shortcoming.

c) Sequencing errors: since Pool-seq relies on NGS technologies which have a relatively higher error rate, it is difficult to distinguish between rare alleles and sequencing errors. Replication of pools can reduce the error rate in SNP-calling.

d) Differential representation of individuals in the pool: this may arise from technical errors in pipetting and DNA quantification.

e) Misalignment: the problem with aligning short reads with a reference genome is that they may *get al*igned to a different genomic location or may not align at all, if the alleles are too divergent. This issue can be resolved by increasing the depth of sequencing.

Czech *et al.* (2022) employed Pool-seq as a methodology for investigating the impacts of natural selection and genetic drift on Arabidopsis populations by observing alterations in allele frequencies. To effectively monitor the chronological evolution of genetic variations, they analyzed extensive experimental or wild populations by grouping individuals sampled across various time points. This was achieved using high-throughput genome sequencing, a technique referred to as the Evolve and Re-sequence approach (E and R).

Best of Both Worlds: Integrating more than one Forms of BSA

Attempts to integrate more than one modification of BSA to increase efficiency and reduce time have also been successful. In tomato, conventional QTL analysis was complemented with QTL-seq to identify QTLs governing heat-tolerance in tomato. This was further supplemented by RNA-Seq identify high-temperature stress-responsive (HSR) genes within the major QTLs in tomato (Wen *et al.*, 2019). QTLs located using conventional QTL analysis were consistent with QTL-seq. RNA-seq identified four candidate genes within the major QTLs by DEG analysis, qRT-PCR screening and biological function analysis. F_2 population was used for bulking.

Salt stress is a significant abiotic factor that negatively affects rice productivity on a global scale. The ability to tolerate salt stress during the bud burst stage is vital for the survival of rice seedlings and the ultimate yield, particularly in the DSR model. Researchers utilized an $F_{2:3}$ population created by crossing IR36 (which is sensitive to salt) with Weiguo (known for salt tolerance) to perform QTL-seq and map quantitative trait loci (QTLs). Through the utilization of ΔSNP-index algorithms and Euclidean distance algorithms, a major QTL was pinpointed on chromosome 7. Subsequent development of KASP markers adjacent to the identified QTL facilitated the narrowing down of the region. By employing a subset of the initial $F_{2:3}$ population, regional QTL mapping was conducted, effectively confining the QTL to a 222 kb genomic segment. To determine the candidate gene, RNA-seq was applied to identify five differentially expressed genes transcribed from this specific genomic locus (Lei *et al.*, 2020).

Ye *et al.* (2022) combined BSA-Seq and RNA-seq to unravel candidate genes associated with plant architecture in *Brassica napus.* A BC8 population was used for bulking. BSA-Seq identified five candidate regions in the B. napus genome. RNA-Seq analysis revealed 4378 DEGs, which were then filtered for genes related to cell structure, plant hormone biosynthesis and signal transduction and the phenylpropanoid pathway. Utilizing association analysis of BSA-Seq and RNA-Seq, it was indicated that a selection of seven differentially expressed genes (DEGs) associated with plant hormone signal transduction, alongside a WUSCHEL-related homeobox (WOX) gene labeled as BnaA01g01910D, could potentially serve as candidate genes attributed to the observed dwarf and compact phenotype evident in one of the parental organisms.

MutMap and its Variants

MutMap, an approach aimed at accelerating the identification of mutant genes and understanding their functional roles, serves as a powerful tool for pinpointing the specific mutation responsible for a particular trait. This method harnesses the advantages of both Bulked Segregant Analysis and whole-genome resequencing techniques (Manchikatla *et al.*, 2021).

To employ MutMap, researchers initiate the process by inducing mutations in the genetic material through mutagenic treatments. The resulting mutant population is then scrutinized to identify individuals displaying the desired trait. Subsequently, the mutant phenotype of interest is crossed with the parental line, giving rise to F_1 and F_2 populations. Utilizing single nucleotide polymorphisms (SNPs), researchers conduct segregation analysis and map the trait's genetic location on the genome. This approach essentially follows the principles of forward genetics (Tribhuvan *et al.*, 2018). It is a method based on whole-genome resequencing of pooled DNA from a segregating population of plants that show a useful phenotype.

Sequence coverage or read depth, in DNA sequencing, is the number of unique reads that include a given nucleotide in the reconstructed sequence. Higher read depth indicates higher confidence in SNP/base calling. The changes in the mutant bulk would be observed in the form of SNP changes. Majority of SNPs will segregate in a 1:1 mutant/wild type ratio. The SNP responsible for the change of phenotype is homozygous in the progeny showing the mutant phenotype. SNPs that are unlinked to the SNP responsible for the mutant phenotype wold show 50 per cent mutant and 50 per cent wild-type. Causal SNP and closely linked SNPs would show 100 per cent mutant and 0 per cent wild-type reads. SNPs loosely linked to the causal mutation = >50 per cent mutant and <50 per cent wild-type reads.

These SNP changes are further indicated in the form of SNP index.

$$\text{SNP index} = \frac{\text{Number of reads of a mutant SNP}}{\text{Total number of reads corresponding to the SNP}}$$

- ✰ Expected SNP index :
    - = 1 near the causal gene
    - = 0.5 for the unlinked loci
- ✰ Calculation is done using sliding window approach (moving averages).

Abe *et al.* (2012) was the first to apply MutMap to mutants of a Japanese rice variety and identified the unique probable genomic positions harboring mutations causing pale green leaf colour and semi-dwarf trait. The mutant stock comprised large number of EMS–mutagenized lines of M_3–M_4 generations with a Hitomebore background. MutMap was applied to a mutant - Hit1917-pl1 characterised by pale-green leaf phenotypes with lesser chlorophyll concentrations compared with wild type. This Mutant x Hitomebore wild type gave F1 progeny which were self-pollinated to get >200 F_2 progeny. The observed segregation ratio between wild-type and mutant phenotypes in field-grown F_2 progeny = 3:1, indicating that each mutant phenotype was caused by a recessive mutation in a single locus.

DNA was bulked from mutant progeny and subjected to whole-genome sequencing and alignment to a reference sequence of Hitomebore, it was found that the mutants harbor 1,001 (Hit1917-pl1) transition-type (G→A and C→T) SNPs. SNP index Manhattan plots for all the 12 chromosomes were constructed and was observed that SNP indices were distributed randomly around 0.5 for most parts of the genome for the two mutant lines. A single unique genomic region showing a cluster of SNPs with SNP index of 1 was observed.

Two out of 7 SNPs in the cluster showed SNP index=1; transformation and Real-time quantitative RT-PCR analysis further confirmed that the Hit1917-pl1 phenotype is caused by the mutation SNP-22981826 as identified by MutMap.

In rice, this method has been applied to map many genes by many researchers such as CAO1, Starch branching enzyme IIb (BEIIb) gene, AGPL2, 08SG2/OsBAK1, OsEDR1 gene, DEP2-1388, LOC_Os06g29380 for leaf colour, grain texture, floury endosperm, grain size and number, spotted-leaf phenotypes, panicle architecture and leaf senescence, respectively by Abe *et al.* (2012), Nakata *et al.* (2018), Zhang *et al.* (2023), Yuan *et al.* (2017), Han *et al.* (2017), Hu *et al.* (2016), Deng *et al.* (2017).

The maize genome, spanning 2300 Mb, is characterized by its extensive complexity, featuring repetitive and highly variable segments. Traits such as kernel row number, plant height, root length, flowering duration, nitrogen use efficiency, as well as responses to abiotic and biotic stresses, demand an isogenic foundation to ensure accurate classification. However, the presence of background modifiers and the significant and pervasive impact of heterosis introduce challenges in precisely evaluating F_2 phenotypes in maize. Constructing a reliable mapping population requires multiple cycles of backcrossing and a comprehensive analysis of pedigrees and trait segregation, processes that demand substantial investments of time and resources.

To facilitate genetic research in a non-standard maize variety, a rapid mutant mapping method was devised by Quan *et al.* (2020). This enhanced Mutmap pipeline, implemented in maize, obviated the need for crossing with an unrelated inbred. The approach involved two generations of crossing the mutant with a mutagenized ML10 line, thereby introducing more segregating variants for mapping. Applied to the moderate fasciation mutant E1-9, the mapping unveiled that the ear tip exhibited mild fasciation and the ear had increased kernel rows compared to the wild type. In the E1-9 and F_2 mutant-pools, a noteworthy 376 bp deletion (Chr4) was identified within the CLE7 promoter. The correlation between the mutant phenotype and this promoter loss was evident through co-segregation, and PCR validation robustly confirmed E1-9 as a CLE7 mutant.

Significantly, the EMS-induced allele's origin from an elite parental line implies its direct applicability in breeding programs to enhance kernel row number (KRN) without requiring transgenic regulation. In summary, this approach offers an expedient strategy to map and exploit advantageous genetic variations for maize improvement.

A MutMap population for early flowering and large seed size was developed in chickpea. Based on the phenotyping of MutMap population, extreme bulks for days to flowering and 100-seed weight were sequenced and aligned and finally, along with candidate gene identification, two markers, for early flowering and 100-seed weight, could be developed and validated using the candidate SNPs (Manchikatla *et al.*, 2021).

MutMap+

A variant of MutMap for cases where generating prognies by crossing is difficult

It is a modified version of MutMap which can be utilized in cases of developmental sterility/lethality. It can be applied by identifying segregating progeny in M_2 generation and mutants can be separately bulked and sequenced. MutMap+ circumvents the need for both crossing and linkage analysis, it can be implemented over a shorter time span. It can also be applied to rapidly identify the genes involved in the control of early plant development and fertility.

Application of MutMap+ for causal SNP identification

The rice mutant Hit9188 was isolated from a rice mutant stock and characterized by dwarfism and pale green leaves + premature death 3 weeks after germination. Therefore, the mutant could not be crossed for normal MutMap analysis. So, the wild-type M_2 plants were allowed to self-fertilize giving rise to the M3 lines. Further, through SNP indexing and confirmation through RNAi, it could be concluded that the mutation detected in OsNAP6 is responsible for the Hit9188 developmental phenotypes.

MutMap+ extends beyond lethality and sterility genes. It is versatile for crucial agronomic traits, especially in species with crossing challenges. Unlike traditional methods needing complex crosses, selfing, and linkage analysis, MutMap+ offers simpler gene isolation. This is particularly advantageous for species where initiating artificial crosses is tedious.

MutMap-Gap

To fill up the gap in the reference genome sequence and find mutations lying beneath

The parental 'reference genome sequence' must contain the genomic segment spanning the causal mutation in order for MutMap to function. The 'reference genome' of a representative cultivar or line of the species, such as the cv. Nipponbare in rice (Oryza sativa), is used to build the parental line 'reference sequence', however this is not a guarantee. The parental line genome is resequenced and aligned to the species-specific reference genome that is made publicly available. At all the single nucleotide polymorphism (SNP) sites discovered between the two lines, the parental line's nucleotides are subsequently substituted with those of the 'reference genome'. Because of this, conventional MutMap analysis cannot find mutations in missing genomic regions (gaps) particular to the parental line. MutMap-Gap is an extended version of MutMap which can be used to identify mutations in such gap regions. It is basically a combination of MutMap and targeted de novo assembly of genomic gap regions.

The MutMap analysis applied to Hit6780 identified a putative genomic region that most likely harbours the causing mutation in the interval from 7.18 to 13.05 Mb on rice chromosome 9, the same location where the causal mutation of Hit5948 was mapped. Although the region had ten SNPs with an SNP index of one, none of them reflected nonsynonymous changes, and no SNP was found in Os09t0327600-01, the candidate gene for Hit5948. As a result, it was hypothesised that the Hit6780 causal mutation is located in the Hitomebore-specific chromosomal area. To discover the causal mutation of Hit6780, MutMap-Gap analysis was successfully utilized by Takagi *et al.*, 2013.

Standard Prerequisite Tools for MutMap Analysis

1. Perl (v5.8.8)
2. R (version 2.15.0)
3. BWA (version 0.5.9-r16)
4. SAMtools (0.1.8 or before)
5. FASTX-To

The MutMap approach, built upon Bulked Segregant Analysis (BSA) and harnessing cutting-edge sequencing technologies, holds tremendous potential for the identification of causative genes underlying specific traits. This method offers a powerful avenue for pinpointing the genetic factors responsible for observed phenotypes. By identifying Single Nucleotide Polymorphisms (SNPs) in regions associated with the phenotypic variations of interest, these SNPs can serve as valuable DNA markers for the process of marker-aided selection.

Importantly, when mutagenesis is carried out on an already high-performing crop cultivar, the MutMap technique becomes even more valuable. It enables breeders to generate a collection of mutants, each accompanied by associated SNP markers. This wealth of genetic resources provides breeders with the means to efficiently create new and improved crop varieties, leveraging the known genomic locations linked to desired traits.

In the evolving landscape of genomics and crop improvement, MutMap has not remained stagnant. The limitations and challenges initially associated with the technique can now be addressed effectively. The expanding availability of complete genome sequences for various crop species has further enriched the application of MutMap. Additionally, the utilization of near isogenic lines has facilitated even finer resolution in identifying phenotype-associated loci.

Furthermore, the MutMap technology has undergone various modifications and adaptations, bolstered by advancements in BSA and MutMap methodologies. This dynamic progress is fueled by the ever-growing availability of genetic resources and tools, creating a fertile ground for innovation and refinement in the realm of plant breeding. As MutMap techniques continue to evolve, they promise to play an instrumental role in accelerating the development of improved crop varieties, capitalizing on the synergy between genetics, genomics, and breeding expertise.

Table 11.1: MutMap Used in some Crops other than Rice

Crop	Mapped Gene	Trait	References
Quinoa	CqCYP76AD1-1	Green hypocotyl mutant (ghy)	Imamura *et al.* (2018)
Cucumber	CLAVATA1-type receptor-like kinase CsCLAVATA1	Dwarfism	Xu *et al.* (2018)
	Cytochrome P450 Gene CsCYP85A1	Dwarf/compactness in plant architecture	Wang *et al.* (2017)
Sorghum	Sobic.002G221000 (Ms9)	Nuclear male sterility	Chen *et al.* (2019)
	Sobic.001G228100 (GDSL-like lipase/ acylhydrolase)	Devoid of epi-cuticular wax (EW)	Jiao *et al.* (2018)
Wheat	MS1	Male Sterility 1 (Ms1)	Wang *et al.* (2017)
Soybean	Glyma.04g242300	Spotted leaf-1 (spl-1)	Amin *et al.* (2019)
Oilseed rape (*Brassica napus*)	Bna.IAA7.C05	Dwarfism	Cheng *et al.* (2019)

REFERENCES

Abe, A., Kosugi, S., Yoshida, K., Natsume, S., Takagi, H., Kanzaki, H., Matsumura, H., Yoshida, K., Mitsuoka, C., Tamiru, M., Innan, H., Cano, L., Kamoun, S., and Terauchi, R. (2012). Genome sequencing reveals agronomically important loci in rice using MutMap. *Nature Biotechnology*, 30(2), 174.

Amin, G. M., Kong, K., Sharmin, R. A., Kong, J., Bhat, J. A., and Zhao, T. (2019). Characterization and Rapid Gene-Mapping of Leaf Lesion Mimic Phenotype of spl-1 Mutant in Soybean (*Glycine max* (*L.*) *Merr.*). *International Journal of Molecular Sciences*, 20(9), 2193. https://doi.org/10.3390/ijms20092193

Bhat, J. A., and Yu, D. (2021). High-throughput NGS-based genotyping and phenotyping: Role in genomics-assisted breeding for soybean improvement. *Legume Science*, 3(3), e81.

Chen, J., Jiao, Y., Laza, H., Payton, P., Ware, D., and Xin, Z. (2019). Identification of the First Nuclear Male Sterility Gene (Male-sterile 9) in Sorghum. *The Plant Genome*, 12, 190020. https://doi.org/10.3835/plantgenome2019.03.0020

Cheng, H., Jin, F., Zaman, Q. U., *et al.* (2019). Identification of Bna.IAA7.C05 as allelic gene for dwarf mutant generated from tissue culture in oilseed rape. *BMC Plant Biology*, 19, 500. https://doi.org/10.1186/s12870-019-2094-2

Czech, L., Peng, Y., Spence, J. P., Lang, P. L., Bellagio, T., Hildebrandt, J.,. Exposito-Alonso, M. (2022). Monitoring rapid evolution of plant populations at scale with Pool-Sequencing. *BioRxiv*, 2022-02.

Darvasi, A., and Soller, M. (1994). Selective DNA pooling for determination of linkage between a molecular marker and a quantitative trait locus. *Genetics*, 138(4), 1365-1373.

Deng, L., Qin, P., Liu, Z., *et al.* (2017). Characterization and fine mapping of a novel premature leaf senescence mutant yellow leaf and dwarf-1 in rice. *Plant Physiology and Biochemistry*, 111, 50–58. https://doi.org/10.1016/j.plaphy.2016.11.012.

Edae, E. A., and Rouse, M. N. (2019). Bulked segregant analysis RNA-seq (BSR-Seq) validated a stem resistance locus in *Aegilops umbellulata*, a wild relative of wheat. *PLoS One*, 14(9), e0215492.

Fekih, R., Takagi, H., Tamiru, M., Abe, A., Natsume, S., Yaegashi, H., Sharma, S., Sharma, S., Kanzaki, H., Matsumura, H., Saitoh, H., Mitsuoka, C., Utsushi, H., Uemura, A., Kanzaki, E., Kosugi, S., Yoshida, K., Cano, L., Kamoun, S., and Terauchi, R. (2013). MutMap+: genetic mapping and mutant identification without crossing in rice. *PLOS One*, 8(7), e68529.

Giovannoni, J. J., Wing, R. A., Ganal, M. W., and Tanksley, S. D. (1991). Isolation of molecular markers from specific chromosomal intervals using DNA pools from existing mapping populations. *Nucleic Acids Research*, 19, 6553–6558. doi: 10.1093/nar/19.23.6553

Gross, J. B., Furterer, A., Carlson, B. M., and Stahl, B. A. (2013). An integrated transcriptome-wide analysis of cave and surface dwelling *Astyanax mexicanus*. *PloS One*, 8(2), e55659.

Han, X., Xu, R., Duan, P., *et al.* (2017). Genetic analysis and identification of candidate genes for two spotted-leaf mutants (spl101 and spl102) in rice. *Chinese Journal of Rice Science*, 39, 346–353. https://doi.org/10.16288/j.yczz.16-416

Hu, Y., Guo, L., Yang, G.,. Wu, Z. (2016). Genetic analysis of dense and erect panicle-2 allele DEP2-1388 and its application in hybrid rice breeding. *Chinese Journal of Crop Science*, 38, 72–81. https://doi.org/10.16288/j.yczz.15-158

Hu, Y., Guo, L., Yang, G., *et al.* (2016). Genetic analysis of dense and erect panicle-2 allele DEP2-1388 and its application in hybrid rice breeding. *Chinese Journal of Crop Science*, 38, 72–81. https://doi.org/10.16288/j.yczz.15-158

Huang, L., Tang, W., Bu, S., and Wu, W. (2020). Brm: A statistical method for QTL mapping based on bulked segregant analysis by deep sequencing. *Bioinformatics*, 36, 2150–2156. doi: 10.1093/bioinformatics/btz861

Huang, L., Tang, W., Bu, S., and Wu, W. (2020). Brm: A statistical method for QTL mapping based on bulked segregant analysis by deep sequencing. *Bioinformatics*, 36, 2150–2156. doi: 10.1093/bioinformatics/btz861

Imamura, T., Takagi, H., Miyazato, A., Ohki, S., Mizukoshi, H., and Mori, M. (2018). Isolation and characterization of the betalain biosynthesis gene involved in hypocotyl pigmentation of the allotetraploid *Chenopodium quinoa*. *Biochemical and Biophysical Research Communications*, 496(2), 280-286. doi: 10.1016/j.bbrc.2018.01.041

Jiao, Y., Burow, G., Gladman, N., Acosta-Martinez, V., Chen, J., Burke, J., Ware, D., and Xin, Z. (2018). Efficient Identification of Causal Mutations through Sequencing of Bulked F_2 from Two Allelic Bloomless Mutants of *Sorghum bicolor*. *Frontiers in Plant Science*, 8, 2267. doi: 10.3389/fpls.2017.02267.

Karczewski, K. J., Dudley, J. T., Kukurba, K. R., Chen, R., Butte, A. J., Montgomery, S. B., and Snyder, M. (2013). Systematic functional regulatory assessment of disease-associated variants. *Proceedings of the National Academy of Sciences*, 110(23), 9607-9612.

Khurana, E., Fu, Y., Colonna, V., Mu, X. J., Kang, H. M., Lappalainen, T.,. and Gerstein, M. (2013). Integrative annotation of variants from 1092 humans: application to cancer genomics. *Science*, 342(6154), 1235587.

Kozak, G. M., Brennan, R. S., Berdan, E. L., Fuller, R. C., and Whitehead, A. (2014). Functional and population genomic divergence within and between two species of killifish adapted to different osmotic niches. *Evolution*, 68(1), 63-80.

Lei, L., Zheng, H., Bi, Y., Yang, L., Liu, H., Wang, J.,. and Zou, D. (2020). Identification of a major QTL and candidate gene analysis of salt tolerance at the bud burst stage in rice (*Oryza sativa* L.) using QTL-Seq and RNA-Seq. *Rice*, 13, 1-14.

Li, L., Li, D., Liu, S., Ma, X., Dietrich, C. R., Hu, H. C.,. and Schnable, P. S. (2013). The maize glossy13 gene, cloned via BSR-Seq and Seq-walking encodes a putative ABC transporter required for the normal accumulation of epicuticular waxes. *PloS One*, 8(12), e82333.

Liu, S., Yeh, C. T., Tang, H. M., Nettleton, D., and Schnable, P. S. (2012). Gene mapping via bulked segregant RNA-Seq (BSR-Seq). *PloS One*, 7(5), e36406.

Majeed, A., Johar, P., Raina, A., Salgotra, R. K., Feng, X., and Bhat, J. A. (2022). Harnessing the potential of Bulked Segregant Analysis sequencing and its related approaches in crop breeding. *Frontiers in Genetics*, 13, 944501.

Manchikatla, P. K., Kalavikatte, D., Mallikarjuna, B. P., Palakurthi, R., Khan, A. W., Jha, U. C., Bajaj, P., Singam, P., Chitikineni, A., Varshney, R. K., and Thudi, M. (2021). MutMap approach enables rapid identification of candidate genes and development of markers associated with early flowering and enhanced seed size in chickpea (*Cicer arietinum* L.). *Frontiers in Plant Science*, 12, p.688694.

Michelmore, R. W., Paran, I., and Kesseli, R. V. (1991). Identification of markers linked to disease-resistance genes by bulked segregant analysis: A rapid method to detect markers in specific genomic regions by using segregating populations. *Proceedings of the National Academy of Sciences of the United States of America*, 88, 9828–9832. doi: 10.1073/pnas.88.21.9828

Neeraja, C. N., Maghirang-Rodriguez, R., Pamplona, A., Heuer, S., Collard, B. C., Septiningsih, E. M.,. and Mackill, D. J. (2007). A marker-assisted backcross approach for developing submergence-tolerant rice cultivars. *Theoretical and Applied Genetics*, 115, 767-776.

Schaub, M. A., Boyle, A. P., Kundaje, A., Batzoglou, S., and Snyder, M. (2012). Linking disease associations with regulatory information in the human genome. *Genome Research*, 22(9), 1748-1759.

Schneeberger, K.; Ossowski, S.; Lanz, C.; Juul, T.; Petersen, A. H.; Nielsen, K. L.; Jorgensen, J. E.; Weigel, D.; Andersen, S. U. (2009). SHOREmap: Simultaneous mapping and mutation identification by deep sequencing. *Nature Methods*, 6, 550–551.

Sloan, D. B., Keller, S. R., Berardi, A. E., Sanderson, B. J., Karpovich, J. F., and Taylor, D. R. (2012). De novo transcriptome assembly and polymorphism detection in the flowering plant *Silene vulgaris* (*Caryophyllaceae*). *Molecular Ecology Resources*, 12(2), 333-343.

Song, J., Li, Z., Liu, Z., Guo, Y., and Qiu, L. J. (2017). Next-generation sequencing from bulked-segregant analysis accelerates the simultaneous identification of two qualitative genes in soybean. *Frontiers in Plant Science*, 8, 919.

Sun, H.; Schneeberger, K. (2015). SHOREmap v3.0: Fast and accurate identification of causal mutations from forward genetic screens. *Methods in Molecular Biology*, 1284, 381–395.

Sun, Y., Wang, J., Crouch, J. H., and Xu, Y. (2010). Efficiency of selective genotyping for genetic analysis of complex traits and potential applications in crop improvement. *Molecular Breeding*, 26, 493-511.

Takagi, H., Abe, A., Yoshida, K., Kosugi, S., Natsume, S., Mitsuoka, C.,. and Terauchi, R. (2013). QTL seq: rapid mapping of quantitative trait loci in rice by whole genome resequencing of DNA from two bulked populations. *The Plant Journal*, 74(1), 174-183.

Takagi, H., Tamiru, M., Abe, A., Yoshida, K., Uemura, A., Yaegashi, H., Obara, T., Oikawa, K., Utsushi, H., Kanzaki, H., Matsumura, H., Urasaki, N., Kamoun, S., and Terauchi, R. (2015). MutMap accelerates breeding of a salt-tolerant rice cultivar. *Nature Biotechnology*, 33(5), 445-449.

Takagi, H., Uemura, A., Yaegashi, H., Tamiru, M., Abe, A., Mitsuoka, C., Utsushi, H., Natsume, S., Kanzaki, H., Matsumura, H., Saitoh, H., Yoshida, K., Cano, L., Kamoun, S., and Terauchi, R. (2013). MutMap Gap: whole genome resequencing of mutant F$_2$ progeny bulk combined with de novo assembly of gap regions identifies the rice blast resistance gene *Pii*. *New Phytologist*, 200(1), 276-283.

Takagi, H., Uemura, A., Yaegashi, H., Tamiru, M., Abe, A., Mitsuoka, C., Utsushi, H., Natsume, S., Kanzaki, H., Matsumura, H., Saitoh, H., Yoshida, K., Cano, L., Kamoun, S., and Terauchi, R. (2013). MutMap Gap: whole genome resequencing of mutant F$_2$ progeny bulk combined with de novo assembly of gap regions identifies the rice blast resistance gene *Pii*. *New Phytologist*, 200(1), 276-283.

Tang, W., Huang, L., Bu, S., Zhang, X., and Wu, W. (2018). Estimation of QTL heritability based on pooled sequencing data. *Bioinformatics*, 34(6), 978-984.

Thakur, V.; Wanchana, S. (2019). Gene Discovery by Forward Genetic Approach in the Era of High-Throughput Sequencing. In R. Banerjee, G. V. Kumar, and S. P. J. Kumar (Eds.), *OMICS-Based Approaches in Plant Biotechnology* (pp. 75–90). Scrivener Publishing LLC.

Tran, Q. H., Bui, N. H., Kappel, C., Dau, N. T. N., Nguyen, L. T., Tran, T. T., Khanh, T. D., Trung, K. H., Lenhard, M., and Vi, S. L. (2020). Mapping-by-Sequencing via MutMap Identifies a Mutation in *ZmCLE7* Underlying Fasciation in a Newly Developed EMS Mutant Population in an Elite Tropical Maize Inbred. *Genes*, 11(3), 281.

Tran, Q. H., Bui, N. H., Kappel, C., Dau, N. T. N., Nguyen, L. T., Tran, T. T.,. Vi, S. L. (2020). Mapping-by-Sequencing via MutMap Identifies a Mutation in *ZmCLE7* Underlying Fasciation in a Newly Developed EMS Mutant Population in an Elite Tropical Maize Inbred. *Genes*, 11(3), 281.

Tribhuvan, K. U., Kumar, K., Sevanthi, A. M., and Gaikwad, K. (2018). MutMap: A versatile tool for identification of mutant loci and mapping of genes. *Indian Journal of Plant Physiology*, 23, 612–621. doi: 10.1007/s40502-018-0417-1

Trick, M., Adamski, N. M., Mugford, S. G., Jiang, C. C., Febrer, M., and Uauy, C. (2012). Combining SNP discovery from next-generation sequencing data with bulked segregant analysis (BSA) to fine-map genes in polyploid wheat. *BMC Plant Biology*, 12(1), 1-17.

Wang, H., Li, W., Qin, Y.,. Ma, L. (2017). The Cytochrome P450 Gene *CsCYP85A1* is a putative candidate for super compact-1 (*Scp-1*) plant architecture mutation in cucumber (*Cucumis sativus* L.). *Frontiers in Plant Science*, 8, 266. https://doi.org/10.3389/fpls.2017.00266

Wang, H., Li, W., Qin, Y.,. Ma, L. (2017). The Cytochrome P450 Gene *CsCYP85A1* is a putative candidate for super compact-1 (*Scp-1*) plant architecture mutation in cucumber (*Cucumis sativus* L.). *Frontiers in Plant Science*, 8, 266. https://doi.org/10.3389/fpls.2017.00266

Wang, Y., Xie, J., Zhang, H., Guo, B., Ning, S., Chen, Y.,. and Liu, Z. (2017). Mapping stripe rust resistance gene *YrZH22* in Chinese wheat cultivar *Zhoumai 22* by bulked segregant RNA-Seq (*BSR-Seq*) and comparative genomics analyses. *Theoretical and Applied Genetics*, 130, 2191-2201.

Wang, Z., Li, J., Chen, S., Heng, Y., Chen, Z., Yang, J.,. and Ma, L. (2017). Poaceae-specific *MS1* encodes a phospholipid-binding protein for male fertility in bread wheat. *Proceedings of the National Academy of Sciences*, 114(47), 12614-12619.

Wen, J., Jiang, F., Weng, Y., Sun, M., Shi, X., Zhou, Y.,. and Wu, Z. (2019). Identification of heat-tolerance QTLs and high-temperature stress-responsive genes through conventional QTL mapping, QTL-seq and RNA-seq in tomato. *BMC Plant Biology*, 19, 1-17.

Wu, Q., Su, Y., Pan, Y. B., Xu, F., Zou, W., Que, B.,. and Que, Y. (2022). Genetic identification of SNP markers and candidate genes associated with sugarcane smut resistance using BSR-Seq. *Frontiers in Plant Science*, 13, 1035266.

Xu, L., Wang, C., Cao, W., *et al.* (2018). *CLAVATA1*-type receptor like kinase *CsCLAVATA1* is a putative candidate gene for dwarf mutation in cucumber. *Molecular Genetics and Genomics*. https://doi.org/10.1007/s00438-018-1467-9.

Ye, S., Yan, L., Ma, X., Chen, Y., Wu, L., Ma, T.,. and Wen, J. (2022). Combined BSA-seq based mapping and RNA-seq profiling reveal candidate genes associated with plant architecture in *Brassica napus*. *International Journal of Molecular Sciences*, 23(5), 2472.

Yuan, H., Fan, S., Huang, J., *et al.* (2017). *08SG2/OsBAK1* regulates grain size and number, and functions differently in Indica and Japonica backgrounds in rice. *Rice*, 10, 25. https://doi.org/10.1186/s12284-017-0165-2.

Zegeye, W. A., Zhang, Y., Cao, L., and Cheng, S. (2018). Whole Genome Resequencing from bulked populations as a rapid QTL and gene identification method in rice. *International Journal of Molecular Sciences*, 19(12), 4000.

Zhang, B., Qi, F., Hu, G., Yang, Y., Zhang, L., Meng, J.,. and Xing, Y. (2021). BSA-seq-based identification of a major additive plant height QTL with an effect equivalent to that of *Semi-dwarf 1* in a large rice F_2 population. *The Crop Journal*, 9(6), 1428-1437.

Zhang, L., You, R., Chen, H., Zhu, J., Lin, L., and Wei, C. (2023). A New SNP in *AGPL2*, Associated with Floury Endosperm in Rice, Is Identified Using a Modified MutMap Method. *Agronomy*, 13(5), 1381.

Zhang, L., You, R., Chen, H., Zhu, J., Lin, L., and Wei, C. (2023). A New SNP in *AGPL2*, Associated with Floury Endosperm in Rice, Is Identified Using a Modified MutMap Method. *Agronomy*, 13(5), 1381.

Zou, C., Wang, P., and Xu, Y. (2016). Bulked sample analysis in genetics, genomics and crop improvement. *Plant Biotechnology Journal*, 14(10), 1941-1955.

Chapter 12

QTL Mapping: Principles and Practical Applications in Plant Breeding

Tushar Arun Mohanty[1], Lavudya Sampath[1], B.J. Antony[1] and Digvijay Singh[2]*

[1]*Ph. D scholar (Genetics and Plant Breeding), Centre for Plant Breeding and Genetics, Tamil Nadu Agricultural University, Coimbatore – 641 003, Tamil Nadu*
[2]*Assistant Professor, Department of Genetics and Plant Breeding, Narayan Institute of Agricultural Sciences, Gopal Narayan Singh University, Jamuhar, Sasaram*
**e-mail: tushararunmohanty@gmail.com*

INTRODUCTION

Quantitative Trait Locus (QTL) mapping is a powerful genetic technique that has revolutionized the field of plant breeding. It serves as a key tool in deciphering the genetic basis of complex traits in plants and provides valuable insights into the underlying mechanisms controlling these traits. The application of QTL mapping has significantly accelerated the development of improved crop varieties with desirable agronomic traits, enhancing agricultural productivity and sustainability. In traditional plant breeding, breeders relied on phenotypic selection, which often resulted in slow progress due to the complex nature of many traits controlled by multiple genes and environmental factors. However, with the advent of molecular markers and advances in genotyping technologies, QTL mapping emerged as a breakthrough method that allowed for the identification and localization of genomic regions associated with quantitative traits.

The fundamental principle behind QTL mapping lies in the detection of statistical associations between genetic markers and the phenotypic variation of traits of interest. These markers are typically molecular tags scattered throughout the genome, reflecting the genetic diversity present within a population. By analyzing the co-segregation of markers and trait phenotypes across a mapping population, breeders can identify regions of the genome, known as QTLs, which influence the expression of quantitative traits. QTL mapping not only facilitates the discovery of genes responsible for specific traits but also provides crucial information about their genetic architecture. Understanding whether traits are governed by a few major genes or numerous minor genes with additive effects is essential for effective plant breeding strategies.

In recent years, QTL mapping has been extensively used in various crops to dissect the genetic basis of economically important traits such as yield, disease resistance, abiotic stress tolerance, nutritional content, and more. Armed with this knowledge, breeders can employ marker-assisted selection (MAS) or marker-assisted breeding (MAB) to efficiently incorporate favorable QTLs into elite breeding lines, thereby accelerating the development of superior crop varieties.

This chapter delves into the principles and practical applications of QTL mapping in plant breeding. It explores the genetic basis of quantitative traits, various experimental designs, molecular markers used in QTL mapping, and linkage and association mapping techniques. Additionally, it discusses case studies highlighting successful QTL mapping efforts and the challenges faced in the process. Furthermore, the chapter delves into the integration of QTLs into breeding programs and emerging technologies that promise to enhance the efficiency and accuracy of QTL mapping.

Overall, QTL mapping serves as a vital bridge between genomics and plant breeding, empowering breeders with the knowledge and tools necessary to unlock the genetic potential of crops, address food security challenges, and create resilient and high-yielding varieties capable of meeting the demands of a rapidly evolving world.

Quantitative Trait Loci

According to Asins (2002), the premise of quantitative trait loci (QTL) detection was derived from the work of Sax (1923). Geldermann (1975) invented the abbreviation QTL, which was evaluated by Slate (2005). A QTL is a region of DNA that influences a characteristic that is measured on a linear (continuous) scale. Hundreds or possibly thousands of such QTLs influence quantitative trait expression (Mackay *et al.*, 2009).On the basis of DNA markers positioned on a linkage map, QTLs are allotted on a chromosome in the vicinity where the statistical probability is significant. As DNA markers are not affected by the environment, after detecting their polymorphism, these can be used as a tool in mapping QTL. Quantitative traits can have varying phenotypic concentration depending upon the allelic diversity at a QTL region, and the functional markers found associated with these QTLs established the importance of QTL analysis. Thus, polygenic traits that could hardly be analyzed by the utilization of customary breeding methods could easily be labeled with DNA markers.

Principle of QTL Analysis

QTL analysis is based on the idea that genes and markers that segregate during meiosis must be transferred together from parent to offspring if they are strongly related (Collard *et al.*, 2005). Because a quantitative trait involves the simultaneous expression of several genes, there must be an area or locus (QTL) that, if coupled with markers, may be analysed for further advantages. The establishment of a segregating population initially, followed by the discovery of marker-trait correlation using genetic and phenotypic profiles, is fundamental components of QTL analysis.

Genetic Basis of Quantitative Traits

The genetic basis of quantitative traits refers to the genetic factors that contribute to the variation of traits with continuous or quantitative variation, such as height, weight, yield, disease resistance, and other complex traits in organisms. Unlike Mendelian traits with discrete phenotypes, quantitative traits are influenced by multiple genes, as well as environmental factors, leading to a range of phenotypic values within a population.

The key aspects of the genetic basis of quantitative traits are as follows:

Polygenic Inheritance

Quantitative traits are typically controlled by multiple genes, each contributing to a portion of the phenotypic variation. These genes are known as polygenes. Each polygene typically has a small effect on the trait, and their combined action results in the continuous range of phenotypic values observed in a population.

Additive and Non-Additive Effects

The genetic effects of polygenes can be broadly categorized into additive and non-additive effects. Additive effects occur when each allele at a polygenic locus contributes independently to the trait variation. In contrast, non-additive effects, such as dominance and epistasis, occur when interactions between alleles at different loci influence the trait.

1. Environmental Influence

Environmental factors play a significant role in determining the final expression of quantitative traits. Environmental conditions, such as temperature, humidity, nutrient availability, and other external factors, can interact with the genetic factors and modulate the trait expression.

2. Phenotypic Plasticity

Many quantitative traits exhibit phenotypic plasticity, meaning they can be influenced by environmental cues during development. Phenotypic plasticity allows organisms to adjust their phenotype to optimize their fitness in varying environmental conditions.

3. Heritability

Heritability measures the proportion of phenotypic variation in a population that can be attributed to genetic variation. High heritability suggests that a substantial portion of trait variation is due to genetic factors, making it more amenable to selection in breeding programs.

4. Genetic Mapping

QTL mapping, as mentioned earlier, is a valuable tool to study the genetic basis of quantitative traits. It involves identifying genomic regions, known as quantitative trait loci (QTLs) that harbor genes contributing to the variation of the trait of interest.

5. GxE Interactions

Gene-by-environment (GxE) interactions occur when the effects of genes on a trait are dependent on the specific environmental conditions. GxE interactions add another layer of complexity to the genetic basis of quantitative traits.

Understanding the genetic basis of quantitative traits is essential for plant and animal breeding, as well as understanding the genetic basis of complex human traits and diseases. Knowledge of the underlying genetic factors and their interactions allows breeders to implement effective breeding strategies, such as marker-assisted selection (MAS), genomic selection, and genome editing, to improve the performance of plants and animals and contribute to the sustainable development of agriculture and food production.

Methodology Involved

The science of quantitative genetics has been predominantly occupied by biometric mathematics. Sophisticated statistical tools are involved to extract and correlate the variation in genotypic and phenotypic diversity among individuals. QTL analysis can be performed if we have the following information:

1. A model segregating the population in which a QTL for the required trait is to be detected.
2. Genetic dissection of the population with markers.
3. A record of phenotypic variations for the trait of interest.
4. Software packages to depict marker–trait association.

Mapping Population

Breeding populations vary from wild populations in that they are selected based on the interests of the breeders. The genetic features of the breeding population are carefully limited and concentrated on needed qualities. All breeding professions follow a consistent pattern of producing new genotypic variants. Crossing the lines with essential characteristics increases the likelihood of identifying a QTL (Würschum 2012). Parent selection is critical for studying the segregation of any polygenic trait of interest. P1 (disease resistant) and P2 (disease susceptible) parents must be phenotypically examined and exhibit difference in trait expression. The mapping population in self-pollinated species should be started from highly homozygous (inbred) parents (Collard *et al.*, 2005). In cross-pollinated species, the F_1 generation can be developed by pair-crossing of heterozygous parent plants that are significantly different for required traits (Barrett *et al.*, 2004). F_2 populations from F_1 hybrids, backcross-derived lines, are the usual types most often programmed for self-pollinated species and can easily be developed in a short time. It is also possible to create recombinant inbred lines (RILs) and doubled haploid (DH) lines. If homozygous lines are to be increased without any specific genetic changes, RILs and DH lines are utilised (Collard *et al.*, 2005). The establishment of a mapping population in QTL mapping must include a technique for establishing a connection between the strength of linkage and the degree of linkage disequilibrium (Gardner and Latta 2007). When an allele at locus A is nonrandomly connected with the allele at locus B, linkage disequilibrium (LD) occurs. It can happen if these two loci are not connected (Flint-Garcia *et al.*, 2003)

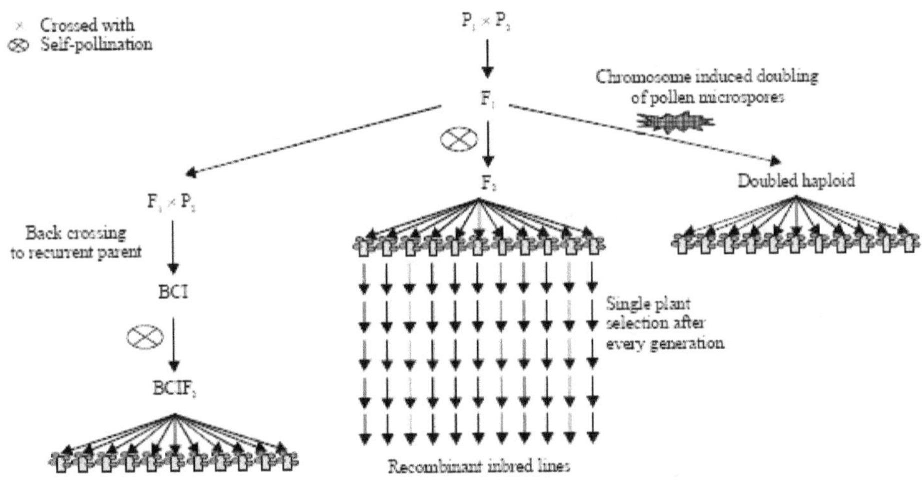

Figure 12.1: Mapping Populations for Self-pollinating Species (Paterson, 1996).

Phenotyping for QTL Mapping

To dissect a characteristic, both the genotypic variation pattern displayed by markers and the phenotypic variety presentation are required. A quantitative characteristic with a continuous distribution pattern is scored, and the whole population as well as the parents are screened. When phenotypic data is not available, this approach is used to detect QTLs. The data is often generated by integrating numerous tests, however the conclusions are unbalanced (Würschum 2012), whereas balanced data sets have been demonstrated to be useful in reducing false-positive QTLs (Wang *et al.,* 2012). Even yet, phenotypic data acquired without a previous balanced experimental design can be utilised to discover QTL.

Trait Selection: Choose the trait of interest carefully, ensuring that it is relevant to the research objectives and has sufficient variation within the mapping population. Traits can include agronomic, morphological, physiological, and biochemical characteristics, as well as disease resistance, yield-related traits, stress tolerance, or any other quantitative trait.

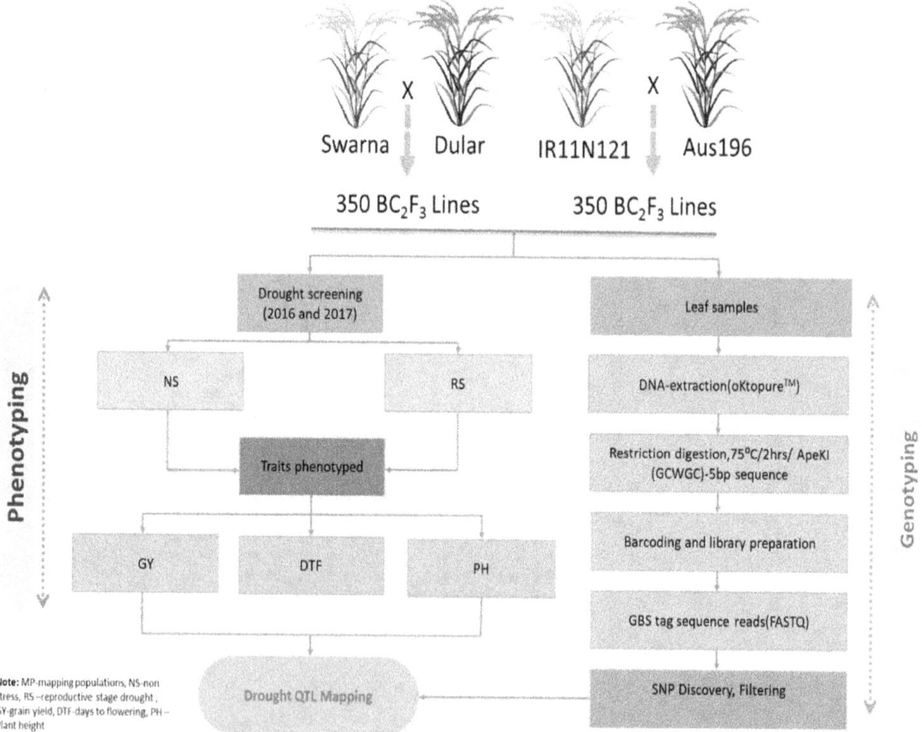

Figure 12.2: Integration of GBS Derived High Density SNPs and Multi-season Phenotyping Data for Mapping of Drought QTLs in Rice (Yadav *et al.*, 2019).

Replication and Controls: Replicates of the mapping population and appropriate control lines or individuals should be included in the phenotyping experiment. Replication helps account for experimental variability and provides statistical power to detect significant QTLs.

Phenotyping Environment: Consider the environmental conditions in which phenotyping will be conducted. Multiple environments or locations (multi-environment trials) may be used to capture genotype-by-environment interactions and identify stable QTLs across diverse conditions.

Experimental Design: Implement a well-designed experimental layout to minimize confounding factors and sources of bias. Randomization and proper experimental controls are essential to ensure the validity of the results.

High-Throughput Phenotyping: For large mapping populations, high-throughput phenotyping platforms and technologies can improve efficiency and reduce time and labor. Automated systems and remote sensing technologies can be used to phenotypically characterize multiple individuals simultaneously.

Data Standardization: Ensure standardized protocols and procedures for data collection to maintain consistency across the phenotyping experiment. Minimize measurement errors and adopt standardized units and scales.

Data Quality Control: Regularly monitor data quality throughout the phenotyping process. Identify and address outliers, inconsistencies, or technical issues that could affect the accuracy of the results.

Statistical Analysis: Choose appropriate statistical methods to analyze the phenotypic data, such as analysis of variance (ANOVA), linear mixed models, or other appropriate models for detecting QTLs.

Integration of Multi-Omics Data: In some cases, integrating phenotypic data with genotypic and other omics data (*e.g.*, transcriptomics, metabolomics) can provide a more comprehensive understanding of the genetic basis of the trait.

Validation: If possible, validate the results of QTL mapping in independent experiments or populations to confirm the presence and effects of the identified QTLs.

Overall, careful and precise phenotyping is essential for successful QTL mapping studies. It ensures that accurate genetic regions associated with quantitative traits are identified, leading to improved crop breeding, enhanced trait understanding, and potential applications in various fields, including agriculture, biotechnology, and medicine.

Genotyping Technologies and Marker Systems

Genotyping technologies and marker systems play a crucial role in QTL mapping for plant breeding. These technologies enable researchers to genotype individuals in mapping populations and identify genetic markers associated with quantitative traits. Here are some of the commonly used genotyping technologies and marker systems in QTL mapping for plant breeding:

Single Nucleotide Polymorphisms (SNPs)

☆ SNPs are the most widely used markers in QTL mapping and plant breeding due to their abundance and cost-effectiveness.

☆ High-throughput genotyping platforms, such as SNP arrays and genotyping-by-sequencing (GBS), enable the simultaneous analysis of thousands to millions of SNPs in a single experiment.

Simple Sequence Repeats (SSRs) or Microsatellites

☆ SSRs are another commonly used marker system in QTL mapping.

☆ SSRs are short, repetitive DNA sequences with high levels of polymorphism, making them valuable for detecting genetic variation within and among populations.

Insertion-Deletion Polymorphisms (InDels)

☆ InDels are small insertions or deletions of DNA segments in the genome.

☆ They are cost-effective markers and are commonly used for genotyping in various plant species.

Diversity Arrays Technology (DArT)

☆ DArT is a marker system that utilizes hybridization-based approaches to genotype individuals.

☆ It allows for high-throughput genotyping of thousands of genomic loci, providing dense marker coverage for QTL mapping.

Restriction Site Associated DNA Sequencing (RAD-Seq)

☆ RAD-Seq is a technique that uses restriction enzymes to selectively sequence genomic regions.

☆ It provides high-density genotyping data and is particularly useful for non-model species without a reference genome.

Genotyping-by-Sequencing (GBS)

☆ GBS is a cost-effective approach that combines reduced representation of the genome with next-generation sequencing.

☆ It allows for the discovery of novel SNPs and provides dense marker coverage for QTL mapping.

High-Throughput Sequencing (HTS) and Whole-Genome Sequencing (WGS)

- ☆ HTS and WGS technologies provide a comprehensive view of the entire genome and enable the identification of all types of genetic variants, including SNPs, InDels, and structural variations.

- ☆ WGS is especially useful for species with well-annotated reference genomes.

The choice of genotyping technology and marker system depends on factors such as the species under study, the number of markers required, available resources, and the desired marker density. The advent of high-throughput genotyping technologies has significantly accelerated QTL mapping studies, allowing researchers to efficiently identify genetic regions associated with important traits for plant breeding and crop improvement.

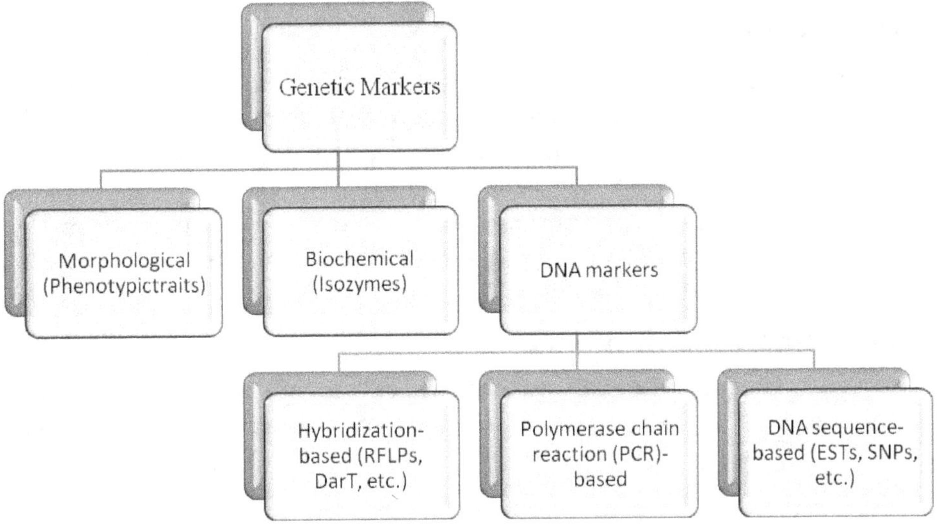

Figure 12.3: Flowsheet Presentation of Genetic Markers.

QTL Mapping Software and Analysis Work Flow

QTL mapping involves complex data analysis, and various software packages are available to facilitate the process. These software packages provide tools for data manipulation, statistical analysis, and visualization of QTL mapping results. Here is a general workflow for QTL mapping analysis using software:

Data Preprocessing

This step involves preparing the data for analysis. It includes data cleaning, formatting, and merging genotype and phenotype data. Genotype data can be in the form of SNP markers or other genetic markers, while phenotype data represents the quantitative trait values.

Data Imputation (Optional)

☆ In some cases, genotype data may have missing values. Data imputation methods can be used to estimate the missing genotypes based on the available information.

Population Structure and Kinship Analysis

☆ Before performing QTL mapping, it is essential to account for population structure and relatedness among individuals in the population.

☆ Principal Component Analysis (PCA) and kinship analysis are commonly used to identify and correct for population structure and relatedness.

QTL Mapping Analysis

☆ QTL mapping is performed using appropriate statistical methods. Commonly used methods include interval mapping, composite interval mapping (CIM), multiple QTL mapping (MQM), and Bayesian approaches like MQM-B.

☆ These methods assess the association between genetic markers and the quantitative trait, taking into account the covariate effects, population structure, and relatedness.

Permutation Tests (Optional)

☆ To determine the statistical significance of identified QTLs, permutation tests can be performed to generate empirical significance thresholds. This helps control false positives.

Visualization of Results

☆ Visualization tools are used to plot the QTL mapping results and identify significant QTL regions.

☆ Manhattan plots and QTL profile plots are commonly used to display marker-trait associations and LOD scores along the genome.

QTL Validation

☆ Identified QTLs should be validated using additional experiments or independent populations to confirm their effects on the trait.

Commonly Used QTL Mapping Software

R/qtl

☆ R/qtl is an extensive R package for QTL mapping analysis.

☆ It provides a wide range of functions for QTL detection in bi-parental, multi-parental, and association mapping populations.

☆ R/qtl also includes tools for data visualization, permutation tests, and QTL validation.

TASSEL (Trait Analysis by aSSociation, Evolution, and Linkage)

☆ TASSEL is a software suite for analyzing complex traits in large germplasm collections.

☆ It offers tools for association mapping, linkage mapping, population structure analysis, and genomic prediction.

☆ TASSEL supports various marker types and population structures.

QTL Cartographer

☆ QTL Cartographer serves the purpose of conducting single-marker regression and interval mapping.

☆ QTL Cartographer is a versatile tool for single-marker regression and interval mapping, compatible with data from F_2, inbred lines, and backcross-derived populations (Luciano *et al.*, 2012).

☆ It produces graphical results and allows for the exploration of various QTL models through simulated data and parameter adjustments, using Rmap input and output files for linkage maps.

☆ It includes multiple QTL mapping methods, such as interval mapping, composite interval mapping, and multiple-QTL mapping.

☆ QTL Cartographer allows visualization of results through genetic maps and LOD score plots.

MapQTL

☆ MapQTL is a user-friendly choice for analyzing experimental populations like BC1, F_2, RIL, DH, and outbred full-sib families in diploid species.

☆ It's ideal for comparing results across different breeding lines, including advanced backcross inbred lines, advanced intermated inbred lines, and doubled haploids from F_2 generations.

☆ Notably, it enables combined analyses of multiple populations with simple experimental designs and automatic marker cofactor selection within a single project.

☆ This software also offers advanced QTL charting and customizable data export tables (Van Ooijen *et al.*, 2000; Van Ooijen 2004; Van Ooijen and Kyazma 2009).

PLABQTL

☆ PLABQTL is a graphical interface tool for QTL mapping analysis.

☆ It is designed for bi-parental populations and provides options for interval mapping and composite interval mapping.

GridQTL

☆ GridQTL is a software tool for QTL mapping in structured populations, such as association panels and nested association mapping (NAM) populations.

☆ It supports both interval mapping and composite interval mapping methods.

GWASpoly

☆ GWASpoly is a tool for genome-wide association studies (GWAS) in autopolyploids, where multiple copies of chromosomes are present.

☆ It allows for QTL mapping in autopolyploid populations.

PLINK

☆ PLINK is a widely used software for genome-wide association analysis in human and model organism populations.

☆ While initially developed for human genetics, it can be adapted for plant breeding association studies.

The choice of software depends on factors such as the type of mapping population, data complexity, and specific analysis requirements. The software should be selected based on its suitability for the data and the specific statistical methods needed for QTL mapping analysis.

Linkage Mapping Methods

Linkage mapping methods are statistical approaches used to identify genetic markers associated with quantitative traits in a mapping population. These methods are based on the principle of co-segregation between genetic markers and the trait of interest. Linkage mapping is commonly used in QTL (Quantitative Trait Locus) mapping studies to identify genomic regions harboring genes that influence complex traits. Here is some of the main linkage mapping methods:

Interval Mapping (IM)

☆ Interval mapping is a widely used method for QTL mapping.

☆ It scans the genome at regular intervals and tests for associations between genetic markers and the trait of interest using statistical tests, such as likelihood ratio tests (LRT) or regression analysis.

☆ IM identifies individual QTLs one at a time and provides estimates of their genetic effects.

Composite Interval Mapping (CIM)

- ✰ CIM improves the precision and accuracy of QTL detection compared to interval mapping.

- ✰ It incorporates additional marker information as covariates to control for the genetic background effects and increase the statistical power to detect QTLs with small effects.

- ✰ CIM is particularly useful when the mapping population is not inbred or is derived from multiple parental lines.

Multiple QTL Mapping (MQM)

- ✰ MQM is an extension of interval mapping that allows the simultaneous detection of multiple QTLs within a mapping population.

- ✰ It accounts for the presence of multiple QTLs and their interactions, making it more suitable for complex traits controlled by multiple genetic factors.

Multi-QTL Mapping with Bayesian (MQM-B)

- ✰ MQM-B is a Bayesian approach to map multiple QTLs and estimate their effects.

- ✰ It uses a Markov Chain Monte Carlo (MCMC) algorithm to sample the posterior distribution of QTL positions and effect sizes, providing a more comprehensive inference of QTLs.

Selective Genotyping

- ✰ Selective genotyping involves genotyping only a subset of individuals in the mapping population.

- ✰ This approach focuses on individuals that exhibit extreme trait values to increase the power to detect QTLs with large effects.

Inclusive Composite Interval Mapping (ICIM)

- ✰ ICIM combines composite interval mapping with a stepwise regression approach to select markers associated with the trait.

- ✰ It improves the efficiency of QTL detection, especially in populations with complex genetic structures.

Non-parametric Approaches

Non-parametric methods, such as Kruskal-Wallis, Mann-Whitney, or permutation tests, do not require assumptions about trait distribution and provide robust QTL mapping results.

The choice of linkage mapping method depends on the specific characteristics of the mapping population, the genetic architecture of the trait, and the computational resources available. These methods have contributed significantly to the discovery of genetic loci associated with important traits in plant breeding and have paved the way for marker-assisted selection and genomic selection approaches for crop improvement.

Significance thresholds and LOD (Logarithm of Odds) score are essential components of linkage mapping methods in QTL mapping. They play a crucial role in determining whether a genetic marker is statistically associated with a quantitative trait and whether it represents a potential QTL. Here's an explanation of their significance:

Significance Thresholds

☆ In linkage mapping, significance thresholds are used to establish a cutoff point for determining the statistical significance of marker-trait associations.

☆ The significance threshold represents the probability level below which a marker is considered to be associated with the trait and is not due to random chance.

☆ Commonly used significance levels include p-values (*e.g.*, $p < 0.05$, $p < 0.01$) and genome-wide error rates, such as the genome-wide significance threshold based on the Bonferroni correction.

LOD Score

☆ The LOD score is a statistical measure used in linkage analysis to assess the strength of evidence for the presence of a QTL.

☆ It compares the likelihood of obtaining the observed marker-trait association under the hypothesis of linkage (QTL present) versus the hypothesis of no linkage (QTL absent).

☆ A LOD score greater than a predefined threshold indicates evidence for the presence of a QTL at that genomic location.

☆ The LOD score is calculated as the logarithm of the odds ratio between the two hypotheses.

Significance thresholds and LOD scores are closely related in linkage mapping. The LOD score is used to assess the statistical significance of a marker-trait association, and the significance threshold is the predefined cut-off value that helps determine whether the LOD score provides strong evidence for the presence of a QTL.

Association Mapping in Plant Breeding

Association mapping, also known as linkage disequilibrium (LD) mapping or genome-wide association studies (GWAS), is a powerful method used in plant breeding to identify genetic markers associated with complex traits in diverse germplasm collections or natural populations. Unlike linkage mapping, which uses controlled crosses and mapping populations, association mapping utilizes the natural variation present in a population to study marker-trait associations. Here's an overview of association mapping in plant breeding:

Natural Diversity in Germplasm

- ☆ Association mapping relies on the genetic diversity present in a population, which can be a diverse set of cultivars, landraces, breeding lines, or wild accessions.
- ☆ The natural variation in the population contributes to differences in the expression of various traits, including quantitative traits of interest.

Marker-Phenotype Association

- ☆ Association mapping identifies statistical associations between genetic markers, such as Single Nucleotide Polymorphisms (SNPs) or Insertion-Deletion Polymorphisms (InDels), and the phenotypic variation of the trait.
- ☆ The markers used in association mapping are often evenly distributed throughout the genome, covering a broad range of genomic regions.

Linkage Disequilibrium

- ☆ Association mapping exploits linkage disequilibrium (LD), which is the non-random association of alleles at different loci due to historical recombination events.
- ☆ LD between markers and causal genes (or quantitative trait loci, QTLs) allows for the identification of marker-trait associations without the need for constructing specific mapping populations.

Population Structure and Kinship

- ☆ Association mapping requires addressing the effects of population structure and relatedness among individuals in the population.
- ☆ Population structure arises from genetic subgroups or ancestry within the population and kinship represents the relatedness among individuals.
- ☆ These factors can lead to spurious marker-trait associations if not properly accounted for during the analysis.

Statistical Analysis

- ☆ Association mapping employs various statistical methods to detect marker-trait associations while controlling for population structure and kinship.
- ☆ Principal Component Analysis (PCA), kinship analysis, mixed linear models (MLM), and other methods are commonly used to minimize false positives and identify significant marker-trait associations.

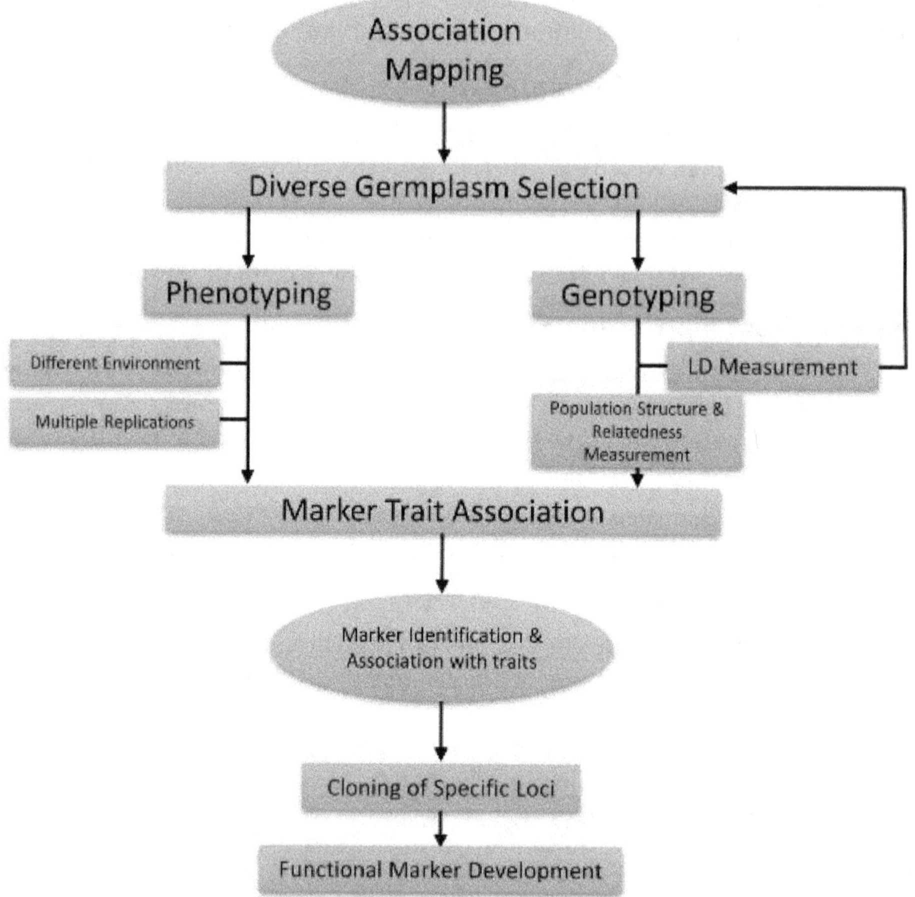

Figure 12.4: The Methodology of Linkage Disequilibrium (LD)-Based Association Mapping (Nadeem *et al.*, 2018).

Marker-Assisted Selection (MAS) and Breeding

☆ The marker-trait associations discovered through association mapping can be utilized in marker-assisted selection (MAS) or genomic selection.

☆ Breeders can use these markers to predict the performance of individuals or lines for specific traits and accelerate the development of improved varieties.

Association mapping has become a valuable tool in plant breeding for identifying candidate genes and markers linked to important agronomic traits, disease resistance, abiotic stress tolerance, and other complex traits. It has significantly expanded our understanding of the genetic basis of traits and accelerated crop improvement efforts, leading to the development of superior varieties with desired traits.

Integrating QTLs into Plant Breeding Programs

Integrating QTLs (Quantitative Trait Loci) into plant breeding programs is a valuable approach to enhance the efficiency and effectiveness of crop improvement efforts. QTLs represent genomic regions associated with important traits, and utilizing this genetic information in breeding can lead to the development of improved crop varieties with desirable traits. Here's how QTLs can be integrated into plant breeding programs:

Marker-Assisted Selection (MAS)

☆ MAS involve using molecular markers linked to QTLs to select individuals with desired trait alleles.

☆ By genotyping breeding lines or individuals at QTL-linked markers, breeders can identify plants carrying favorable alleles for the target trait and use them as parents for the next generation.

Genomic Selection

☆ Genomic selection is a prediction-based breeding strategy that utilizes genome-wide marker information to estimate the breeding value of individuals for multiple traits simultaneously.

☆ QTL information, along with genome-wide marker data, is used to build predictive models that estimate the genetic potential of individuals for various traits.

Marker-Assisted Backcrossing (MABC)

☆ MABC involves introgressing QTLs from a donor parent into elite breeding lines while simultaneously selecting for other desirable traits.

☆ Molecular markers linked to the QTL of interest are used to track the introgression during backcrossing.

Introgression of QTLs through Crossing

☆ By crossing elite lines with donor lines carrying favorable QTL alleles, breeders can introgress the desired QTLs into elite genetic backgrounds.

☆ Recurrent selection and additional breeding cycles can be employed to further enhance the frequency of the favorable QTL alleles.

Multi-Parent Advanced Generation Inter-Cross (MAGIC) Populations

☆ MAGIC populations are designed to capture a diverse range of QTL alleles from multiple parental lines.

☆ The population is used for both QTL mapping and subsequent breeding, enabling the integration of multiple QTLs into breeding lines.

Incorporating Genomic Data into Breeding Decisions

☆ The genetic information derived from QTL mapping studies can guide breeding decisions, such as parent selection and crossing strategies.

☆ Predictive modeling and data analytics can aid breeders in making more informed decisions based on the available genomic data.

High-Throughput Phenotyping and Genotyping

☆ To effectively integrate QTLs into plant breeding, high-throughput phenotyping and genotyping technologies are essential.

☆ These technologies enable breeders to efficiently screen large populations for desired traits and associated QTLs.

By integrating QTL information into plant breeding programs, breeders can accelerate the development of improved crop varieties with enhanced yield, quality, disease resistance, abiotic stress tolerance, and other desirable traits. The incorporation of genomic information into breeding decisions can lead to more targeted and efficient breeding strategies, ultimately benefiting farmers and consumers with better-performing and resilient crops.

Case Studies in QTL Mapping

QTL Mapping for Disease Resistance in Crops

Case Study: Wheat Stem Rust Resistance

☆ Wheat stem rust, caused by the fungus Puccinia graminis f. sp. *tritici*, is a devastating disease that can cause significant yield losses in wheat.

☆ QTL mapping studies have been conducted in wheat to identify genetic loci associated with stem rust resistance.

☆ Researchers have identified several QTLs for stem rust resistance, and some of them have been successfully introgressed into elite wheat cultivars using marker-assisted selection.

☆ The incorporation of these resistance QTLs has contributed to the development of stem rust-resistant wheat varieties, enhancing global food security.

Case Study: Potato Late Blight Resistance

☆ Late blight, caused by the oomycete pathogen Phytophthora infestans, is a devastating disease in potato.

☆ QTL mapping studies in potato have identified QTLs associated with late blight resistance.

☆ Researchers have developed potato varieties with improved late blight resistance by introgressing these resistance QTLs through marker-assisted selection.

☆ The incorporation of late blight resistance QTLs has helped reduce the use of fungicides and increased the resilience of potato crops against this destructive disease.

QTL Mapping for Yield-Related Traits

Case Study: Maize Grain Yield

☆ Grain yield is a complex trait influenced by multiple genetic and environmental factors.

☆ QTL mapping studies in maize have identified QTLs associated with grain yield, kernel number, and other yield-related traits.

☆ Researchers have used these QTLs in breeding programs to develop high-yielding maize hybrids with improved agronomic performance.

☆ The integration of yield-related QTLs has contributed to the continuous improvement of maize varieties, meeting the growing demands for food and feed.

Case Study: Rice Grain Yield and Quality

☆ Grain yield and quality are essential traits in rice breeding for ensuring food security and market value.

☆ QTL mapping studies in rice have identified QTLs associated with grain yield, grain weight, and grain quality attributes like aroma, amylose content, and chalkiness.

☆ Breeders have utilized these yield-related QTLs to develop high-yielding rice varieties with desirable grain quality characteristics, meeting consumer preferences and market demands.

QTL Mapping for Abiotic Stress Tolerance

Case Study: Rice Drought Tolerance

☆ Drought is a major abiotic stress affecting rice production worldwide.

☆ QTL mapping studies in rice have identified QTLs associated with drought tolerance, including root traits, water-use efficiency, and stay-green characteristics.

☆ Breeders have utilized these QTLs to develop drought-tolerant rice varieties that maintain productivity even under water-limited conditions.

☆ The deployment of drought-tolerant rice varieties has helped mitigate the impact of water scarcity on rice production and farmers' livelihoods.

Case Study: Maize Drought Tolerance

- ☆ Drought is a critical abiotic stress affecting maize production in many regions.

- ☆ QTL mapping studies in maize have identified QTLs associated with drought tolerance, including traits related to water-use efficiency, root characteristics, and stay-green traits.

- ☆ Breeders have incorporated these drought tolerance QTLs into maize breeding programs, resulting in the development of drought-tolerant maize hybrids that maintain productivity under water-limited conditions.

QTL Mapping for Heat Tolerance in Wheat

- ☆ Heat stress is a significant concern for wheat production, especially during reproductive stages.

- ☆ QTL mapping studies in wheat have revealed QTLs associated with heat tolerance, including traits related to heat shock protein expression and canopy temperature.

- ☆ These QTLs have been used in breeding programs to develop heat-tolerant wheat varieties that can withstand high-temperature stress, ensuring better grain filling and yield stability.

Advances in QTL Mapping and Future Perspectives

Advances in QTL mapping have been substantial over the years, driven by advancements in genomics, high-throughput phenotyping, and statistical methodologies. These improvements have significantly enhanced our understanding of the genetic basis of complex traits and accelerated crop improvement efforts. Here are some key advances in QTL mapping and future perspectives:

High-Density Genotyping

- ☆ Advancements in genotyping technologies, such as SNP arrays and next-generation sequencing, have enabled the genotyping of large numbers of markers distributed throughout the genome.

- ☆ High-density genotyping facilitates fine-mapping of QTLs and improves the precision in identifying candidate genes associated with the trait of interest.

High-Throughput Phenotyping

- ☆ High-throughput phenotyping platforms allow researchers to accurately and rapidly measure multiple phenotypic traits on a large scale.

☆ Combined with high-density genotyping, high-throughput phenotyping provides more comprehensive data for QTL mapping studies, capturing the phenotypic variation across diverse environments.

Multi-Omics Approaches

☆ Integrating multi-omics data, including genomics, transcriptomics, proteomics, and metabolomics, can provide a deeper understanding of the molecular mechanisms underlying QTL effects.

☆ Multi-omics approaches enhance the identification of candidate genes and pathways associated with complex traits.

Multi-Parent Populations

☆ The use of multi-parent populations, such as Nested Association Mapping (NAM) and Multi-Parent Advanced Generation Inter-Cross (MAGIC) populations, has become more common in QTL mapping.

☆ These populations capture more genetic diversity, allowing for the identification of rare and low-frequency QTL alleles that are not present in bi-parental populations.

Integrative Mapping and GWAS

☆ Integrative mapping methods combine information from multiple mapping populations and genome-wide association studies (GWAS) to improve the power and accuracy of QTL mapping.

☆ These approaches enable the integration of diverse germplasm resources and provide a comprehensive assessment of QTL effects.

Machine Learning and AI

☆ Machine learning and artificial intelligence (AI) techniques are increasingly being used to analyze large-scale genomics and phenomics datasets.

☆ AI-based algorithms can aid in identifying complex relationships between markers and traits, leading to more accurate QTL mapping and prediction models.

Future Perspectives

Advancements in sequencing technologies and statistical methods are enabling precise fine mapping of Quantitative Trait Loci (QTLs) and identification of causal variants, vital for tailored trait manipulation in breeding. Integrating QTL information into genomic selection models accelerates crop variety development, while multi-trait QTL mapping enhances the resilience of crops. Gene editing technologies, like CRISPR-Cas9, enable precise manipulation of QTL-associated genes for trait improvement, and integrating QTL data into systems biology and network approaches enhances our understanding of complex trait interactions, improving predictability and informing breeding strategies.

Limitations and Challenges in QTL Mapping

QTL mapping is a robust method for identifying genomic regions associated with complex traits, but it confronts several limitations and challenges. Firstly, traits in agriculture often result from the interplay of multiple genes and environmental influences, making their dissection complex. The statistical power of QTL mapping relies on sample size and effect sizes, posing difficulties with small populations and subtle QTL effects. Environmental variability introduces genotype-environment interactions, potentially masking QTL effects. Population structure and relatedness can lead to erroneous QTL detections if not properly accounted for. Marker density impacts QTL resolution and linkage disequilibrium can widen confidence intervals. Validation and replication are essential to confirm results, considering potential false positives. Pleiotropy and linkage among QTLs complicate the isolation of individual trait effects. Missing data and imputation introduce uncertainties. Finally, while QTLs point to candidate genes, functional validation necessitates further experiments.

Addressing these limitations and challenges is essential for improving the accuracy and utility of QTL mapping in crop improvement. Integrating multiple data sources, employing advanced statistical methodologies, and leveraging advancements in genomics and phenomics can help overcome some of these limitations and enhance the effectiveness of QTL mapping in breeding programs.

Conclusion

In conclusion, QTL mapping: principles and practical applications in plant breeding has proven to be a powerful approach for identifying genetic loci associated with complex traits in crops. By utilizing high-density genotyping, high-throughput phenotyping, and advanced statistical methods, QTL mapping has significantly advanced our understanding of the genetic basis of agronomic and quality traits. The integration of QTL information in marker-assisted selection and genomic selection has accelerated crop improvement efforts, leading to the development of improved varieties with enhanced yield, stress tolerance, and other desirable traits. While challenges such as trait complexity, environmental variability, and population structure remain, ongoing advancements in genomics and data analytics offer promising avenues for further enhancing the accuracy and efficiency of QTL mapping in breeding programs. Ultimately, QTL mapping continues to be a fundamental tool in modern plant breeding, contributing to sustainable agriculture and addressing global challenges in food production.

REFERENCES

Asins, M. (2002). Present and future of quantitative trait locus analysis in plant breeding. *Plant Breed,* 121, 281–291.

Barrett, B., Griffiths, A., Schreiber, M., Ellison, N., Mercer, C., Bouton, J., Ong, B., Forster, J., Sawbridge, T., and Spangenberg, G. (2004). A microsatellite map of white clover. *Theor Appl Genet,* 109, 596–608.

Collard, B., Jahufer, M., Brouwer, J., and Pang, E. (2005). An introduction to markers, quantitative trait loci (QTL) mapping and marker-assisted selection for crop improvement: the basic concepts. *Euphytica,* 142, 169–196.

Gardner, K. M., and Latta, R. G. (2007). Shared quantitative trait loci underlying the genetic correlation between continuous traits. *Molecular Ecology,* 16(20), 4195-4209.

Geldermann, H. (1975). Investigations on inheritance of quantitative characters in animals by gene markers I. Methods. *Theoretical and Applied Genetics,* 46(7), 319-330.

Luciano Da Costa, E. S., Wang, S., and Zeng, Z-B. (2012). Composite interval mapping and multiple interval mapping: procedures and guidelines for using Windows QTL cartographer. *Methods Mol Biol,* 871, 75–119.

Mackay, T. F., Stone, E. A., and Ayroles, J. F. (2009). The genetics of quantitative traits: challenges and prospects. *Nat Rev Genet,* 10, 565–577.

Nadeem, M. A., Muhammad Amjad Nawaz, M. A., Shahid, M. Q., Doðan, Y., Comertpay, G., Yýldýz, M., Hatipoðlu, R., Ahmad, F., Alsaleh, A., Labhane, N., Özkan, H., Chung, G., and Baloch, F. S. (2018). DNA molecular markers in plant breeding: current status and recent advancements in genomic selection and genome editing. *Biotechnology and Biotechnological Equipment,* 32(2), 261-285.

Paterson, A. H. (1996). Making genetic maps. In: A. H. Paterson (Ed.), Genome Mapping in Plants, 23–39. R. G. Landes Company, San Diego, California; Academic Press, Austin, *Texas Phaseolus vulgaris.*

Sax, K. (1923). The association of size differences with seed-coat pattern and pigmentation in. *Genetics,* 8, 552

Slate, J. (2005). Quantitative trait locus mapping in natural populations: progress, caveats and future directions. *Mol Ecol,* 14, 363–379.

Van Ooijen, J. (2004). MapQTL® 5, software for the mapping of quantitative trait loci in experimental populations. Kyazma, Wageningen, Netherlands.

Van Ooijen, J., Boer, M., Jansen, R., and Maliepaard, C. (2000). MapQTL 4.0: software for the calculation of QTL positions on genetic maps (user manual).

Van Ooijen, J., and Kyazma, B. (2009). MapQTL 6. Software for the mapping of quantitative trait loci in experimental populations of diploid species. Kyazma, Wageningen, Netherlands.

Wang, H., Smith, K. P., Combs, E., Blake, T., Horsley, R. D., and Muehlbauer, G. J. (2012). Effect of population size and unbalanced data sets on QTL detection using genome-wide association mapping in barley breeding germplasm. *Theor Appl Genet*, 124, 111–124.

Würschum, T. (2012). Mapping QTL for agronomic traits in breeding populations. *Theor Appl.*

Yadav, S., Sandhu, N., Singh, V. K., Catolos, M., and Kumar, A. (2019). Genotyping-by-sequencing based QTL mapping for rice grain yield under reproductive stage drought stress tolerance. *Scientific Reports*, 9(1), 14326.

Chapter 13

Site-Directed Mutagenesis in Crop Improvement

Raiza Christina, G.[1], Sumaiya Sulthana, J.[2], Puja Mandal[2] and Aswini M.S.[3]

[1]*Ph.D Scholar, Agricultural College and Research Institute, Madurai, Tamil Nadu*
[2]*Ph.D. Scholar, TNAU, Coimbatore, Tamil Nadu*
[3]*Ph.D. Scholar, College of Agriculture, KAU, Vellayani, Kerala*
**e-mail: raizachrist1997@gmail.com*

INTRODUCTION

A mutation is a heritable, durable modification in the genetic code that can affect how proteins function and alter phenotypic characteristics. The process of producing or creating mutations in an organism's DNA is referred to as "mutagenesis." To research the impact of certain mutations, comprehend gene function, and create new features or characteristics in organisms, this technique is frequently conducted in a regulated manner in labs. Genetics, molecular biology, agriculture, and medicine are a few of the domains where mutagenesis is vital. Spontaneous and induced mutagenesis are the two basic categories of mutation. Errors in DNA replication, DNA repair procedures, or other biological processes can result in spontaneous mutagenesis. Since spontaneous mutation is the primary source of genetic variation, it helps organisms evolve through time. Exposing organisms or cells to agents or situations which enhance the incidence of spontaneous mutations is known as induced mutagenesis. These mutations don't target a particular DNA sequence; instead, they arise randomly over the whole genome. Through the use of external agents known as mutagens, such as chemicals, radiation, and other physical or biological elements, induced mutagenesis deliberately inserts alterations into an organism's DNA. Induced mutagenesis aims to produce a wide variety of mutations so that their impact on various biological processes may be studied, or so that genetic diversity can be produced for evolutionary or breeding research.

Site-directed mutagenesis (SDM) is a form of induced mutagenesis in which DNA may be altered at a particular nucleotide site, changing an intended amino acid. The idea to intentionally alter each DNA sequence in vitro or in vivo was initially put out by Michael Smith and colleagues. For his important contributions to the creation of oligonucleotide-based, site-directed mutagenesis and its use for protein investigations, the late Michael Smith was honoured with the Nobel Prize in 1993. The generation of mutation at one or more places throughout the genome has been accomplished via site-directed mutagenesis. Techniques and enzymes that make these particular deoxyribonucleic acid (DNA) changes easily in vitro in tiny or reasonably large amplicons are no longer the limiting constraint. DNA sequence changes can be done at a single site or several sites within a specific sequence region. Base pairs can be changed, added, or removed. These changes are frequently performed to modify the encoded amino acids or the protein-binding site. The impact of substituting one amino acid with every other amino acid at the same position can be investigated. At the DNA level, this is accomplished by focusing on the coding triplet and changing the nucleotides (degenerate codon, NNN, or NNK) to provide a site-directed repertoire of every probable amino acid encoded by the targeted coding triplet.

There are three different approaches to site-directed mutagenesis: vector-based, PCR-based, and nuclease-based. Plasmid or phage vectors may be utilised in the vector-based mutagenesis process. One mutagenic primer and one normal primer may be employed in this approach of mutation. The process for PCR-based site-directed mutagenesis requires the simultaneous annealing of two oligonucleotide primers, one of which is mutagenic and the other of which is normal, to denatured double-stranded DNA. Site-directed mutagenesis based on nucleases uses enzymes that cleave DNA at a particular sequence. The designed nucleases with greater specificity and efficiency include Zinc Finger Nucleases (ZFNs) and Transcription Activator-Like Effector Nuclease (TALEN). ZFNs and TALENs guide the same non-specific nuclease to break the genome at a particular place using DNA-binding motifs. 2016a, b; Li *et al.*, The most intriguing advancement in gene-editing technology is called Clustered Regularly Interspaced Short Palindromic Repeats (CRISPR). The CRISPR system is an RNA-based bacterial defense system intended to detect and get rid of foreign DNA from bacteriophages and plasmids that are invading cells.

This chapter discusses the methods utilised for these site-directed mutagenesis approaches in detail.

Vector Based Site Directed Mutagenesis

Site-directed mutagenesis using a cloning vector is a technique for introducing certain mutations into a DNA molecule. A tiny fragment of DNA called the cloning vector may be readily modified and spread across bacteria. It has several characteristics that make it ideal for site-directed mutagenesis, such as: 1) selectable marker, such as an antibiotic resistance gene, that enables the bacteria to be chosen for after the mutagenesis operation, 2) The promoter, a DNA sequence that instructs the bacterium to transcribe the DNA molecule, and 3) The polylinker, a section of DNA that is home to several restriction enzyme sites. This enables a precise placement of the DNA molecule into the cloning vector. This approach falls into one of two categories: (1) cassette mutagenesis and (2) enzymatic extension of a mutagenic oligonucleotide annealed to a DNA template.

A restriction fragment from the targeted DNA clone is exchanged for a new restriction fragment that has the required nucleotide alterations in a process. For example, two restriction enzymes, such as XbaI and BglII, are used to cleave a plasmid harbouring a novel clone (YFG; black segment), each of which has just one restriction site throughout the whole plasmid. Agarose gel electrophoresis is used to separate the reaction mixture, and the bigger fragment is then purified from the gel. By using automated DNA synthesis, two single-stranded oligonucleotides are created. The oligonucleotide sequences are complementary to one another and only deviate from the wild-type sequence at the one place (black stripe) that contains the required modifications. The oligonucleotides are combined in a solution that encourages the hybridization of the two strands because of their complementary sequences. Single-stranded, cohesive, sticky ends at the ends of the duplex fragment connect with XbaI and BglII sites. The DNA cassette and the separated fragment are combined, and the T4 DNA ligase joins the two molecules covalently. The ligated DNA is transformed into E. coli, and colonies with antibiotic resistance are chosen. The DNA for plasmids is extracted from distinct bacterial colonies. All colonies have plasmids with the mutant sequence because the two linear segments alone cannot cause E. coli to undergo transformation. Almost all of the clones produced through cassette mutagenesis have the required new (mutant) sequence. The problem with this approach is that it frequently fails to produce distinct, appropriately spaced restriction endonuclease recognition sites to excise a short cassette in the required location.

In Figure 13.2,an oligonucleotide that encodes the desired new DNA sequence is inserted into a cloned DNA fragment. Using a circular, single-stranded, wild-type DNA template, a short oligonucleotide–typically 20–40 nucleotides in length–is hybridised to its complementary sequence. The oligonucleotide's sequence has been designed in a way that the new (mutant) sequence lies in the centre and is bordered on both ends by wild-type sequences.

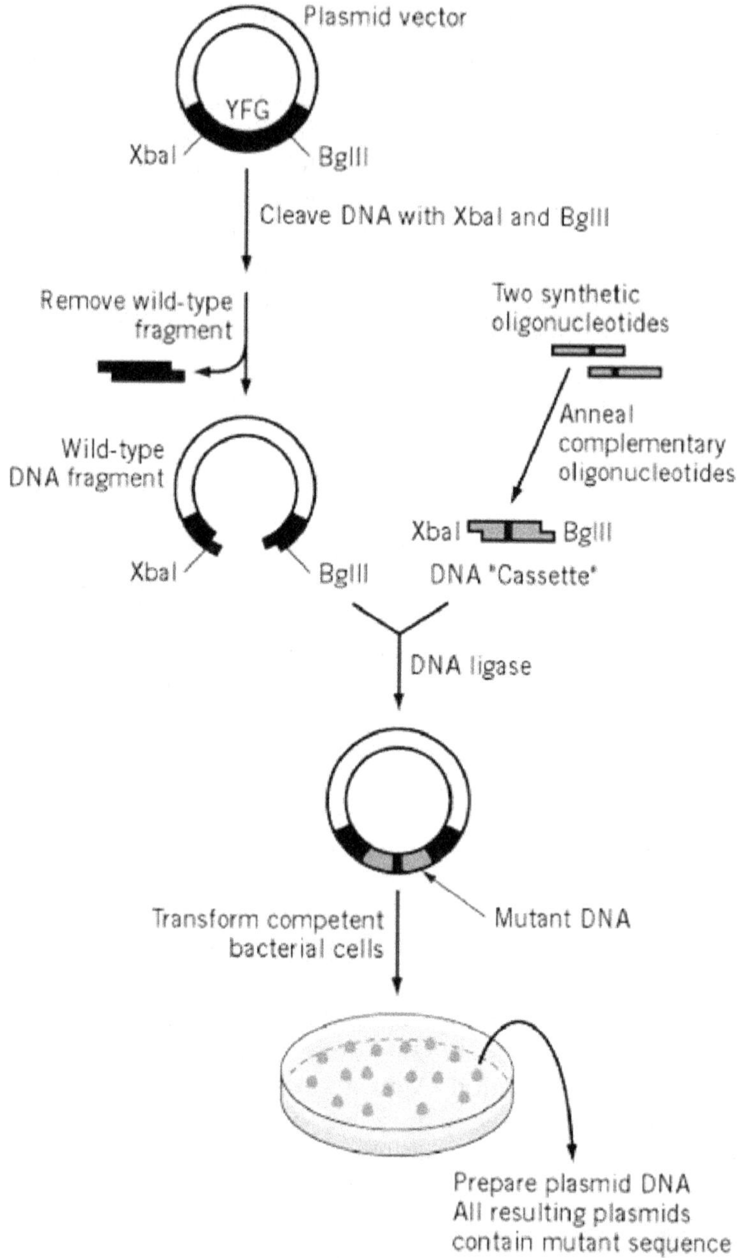

Figure 13.1: Cassette Mutagenesis.

The wild-type sequences in the oligonucleotide are completely complementary to the template DNA when it is annealed to it, but the mutant sequences are mismatched. For the oligonucleotide to successfully create a stable duplex

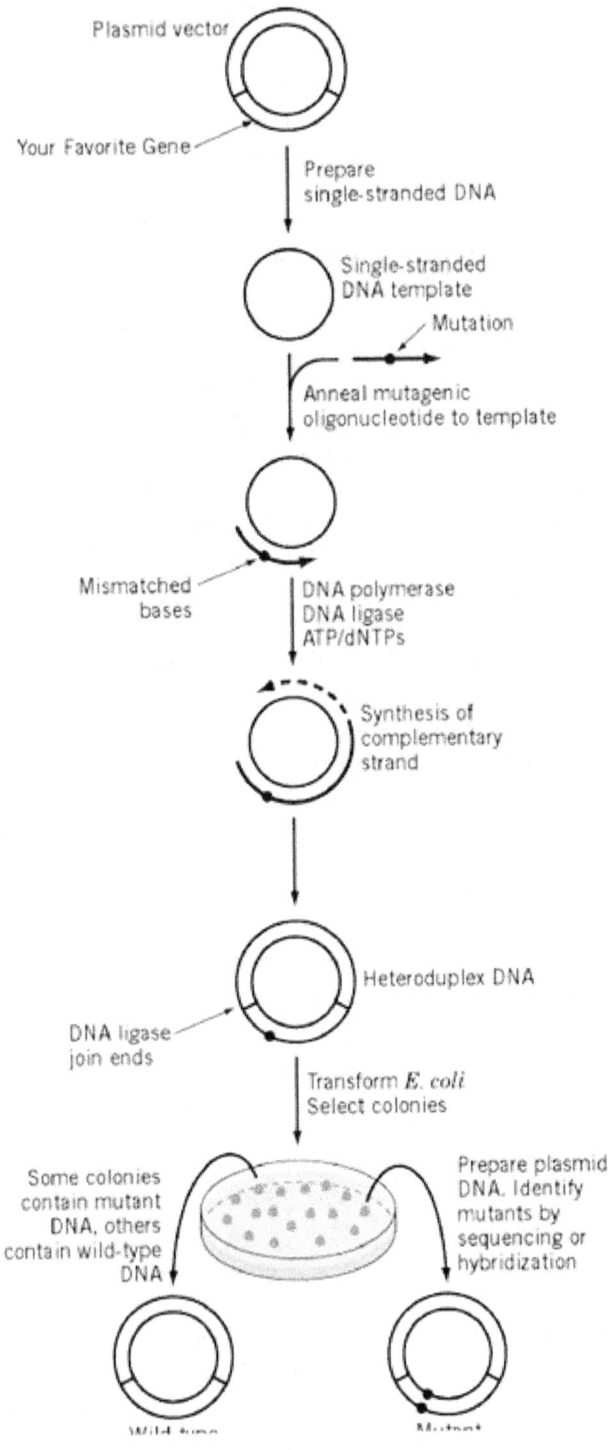

Figure 13.2: Enzymatic Extension of a Mutagenic Oligonucleotide.

with the template DNA, the wild-type sequences must be sufficiently lengthy. The mutagenic oligonucleotide is then used as a primer for DNA synthesis, and nucleotide precursors (dNTPs) are added to synthesise the remaining complementary strand. The mutagenic oligonucleotide is then used as a primer for DNA synthesis, and nucleotide precursors (dNTPs) are added to synthesise the remaining complementary strand. The ends of the freshly synthesised strand are linked enzymatically by T4 DNA ligase once the complementary DNA strand has been entirely wrapped around the circular template. By doing this, a single-stranded DNA molecule is changed into a double-stranded one, with the exception of the area containing the mismatched nucleotides, where all sequences will be wild-type. Competent E. coli cells are then given the resultant double-stranded heteroduplex DNA molecule, which consists of one wild-type strand and one mutant strand. Colonies with plasmid DNA that are drug-resistant are chosen. In the bacterial cells, DNA replication separates the two strands. The bacteria's plasmid DNA is extracted, and clones containing the desired mutation are found by sequencing the plasmid DNA. Theoretically, each colony should include both wild-type and mutant plasmids, but in reality, mismatch repair that occurs before the two strands are replicated sometimes results in just one sequence being retrieved from a colony. In general, mutant plasmids occur less often than the natural type. The most likely cause of this is insufficient DNA replication in vitro. To increase the likelihood of finding plasmids with the desired mutation, many selection or screening techniques have been created. In the past, site-directed mutagenesis was carried out using E. coli DNA polymerase derivatives. The enzyme from E. coli has recently been substituted with DNA polymerases from bacteriophage T4 or T7. The mutagenic oligonucleotide is not replaced or proofread by these DNA polymerases, which are more processing than the E. coli enzyme.

PCR Based Site Directed Mutagenesis

PCR based Site-directed mutagenesis is a technique that uses PCR to introduce certain mutations into a DNA molecule. Small sequence alterations can be introduced into primers by mutations, and larger mutant sections can be produced using primer extension or inverse PCR. Researchers can examine the effects of sequence modifications or screen a variety of mutants using these site-directed mutagenesis approaches to find the best possible sequence to answer the topic at hand. Here, a number of PCR-based site-directed mutagenesis techniques have been reviewed.

Traditional PCR

One of the most common and simple ways for site-directed mutagenesis is the conventional PCR method. The approach is straightforward: the mutation we want to analyses is built into a brief, single-stranded oligonucleotide primer. We can integrate the mutation by using the exonuclease-inactive Taq DNA polymerase. It cannot be eliminated since it is unable to detect the mismatch during amplification. The primer attaches to the DNA and amplifies if the mutation is slight enough to integrate. The DNA sequence changes during the course of an amplification cycle, and the new DNA sequence that we have inserted replaces the old one. For traditional PCR-based mutagenesis, the main guideline is to insert mutant bases up to a certain limit at the 52 end of the primer or in the center of the primer. The Taq polymerase stretches the DNA starting at the 32 end. Therefore, the likelihood of a failed response is increased. The present approach has another drawback in that it contains mixed forms of DNA, including mutant DNA as well as normal/target DNA because template DNA is present.

The primers are created with the intended modification, which might be a base substitution, addition, or deletion, in mind when PCR is used for site-directed mutagenesis. During the PCR process, the mutation replaces the original sequence in the amplicon. Only sections of sequence that are complementary to the primers may be altered by PCR, not the areas between the primers.

For Base Substitution

A non-complementary break in the primer sequence, which will replace the original sequence, is used in PCR primers that include the desired base modifications. The non-complementary bases are added to the amplification product as the primers are extended, replacing the original sequence (Figure 13.3)

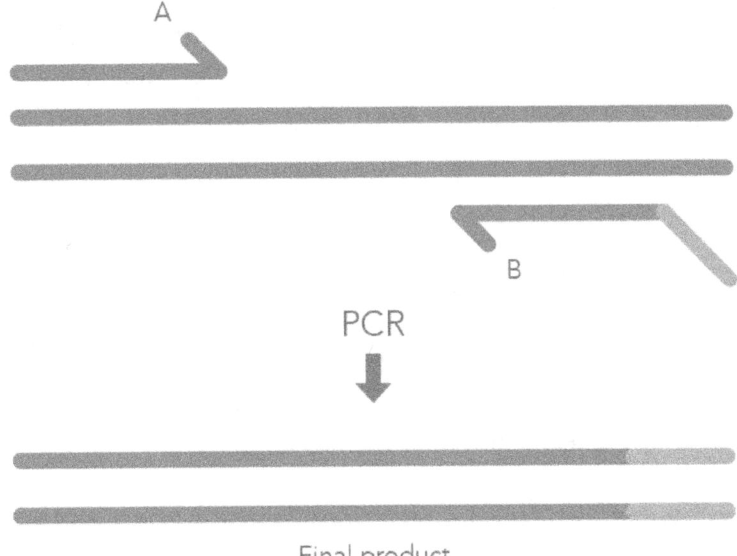

Figure 13.3: PCR for Base Substitution.

For Terminal Addition

Using complementary primer A and an oligonucleotide primer with an addition to the sequence on the 5′ end of primer B, PCR is utilised to create a new PCR product that has the terminal addition (Figure 13.4).

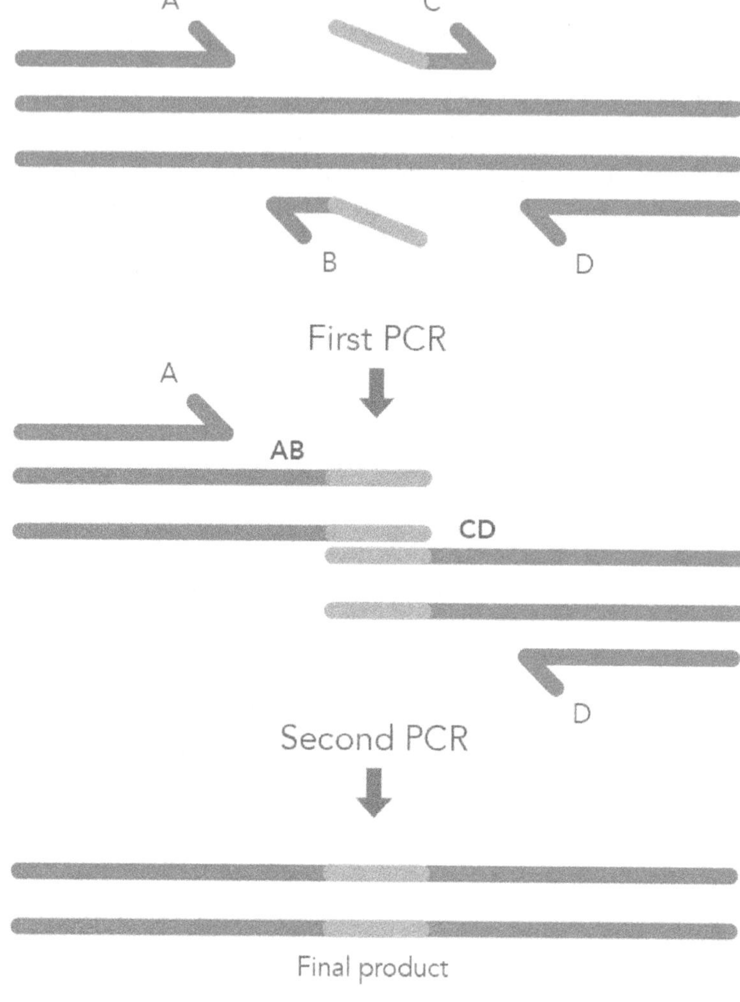

Figure 13.4: PCR for Terminal Addition.

For Deletion

The nucleotides in primer A are complementary to the areas around the region that has to be deleted. A portion of the template will loop out during PCR, amplifying just the complimentary area covered by the primers. The omitted deleted sequence makes the finished product shorter (Figure 13.5).

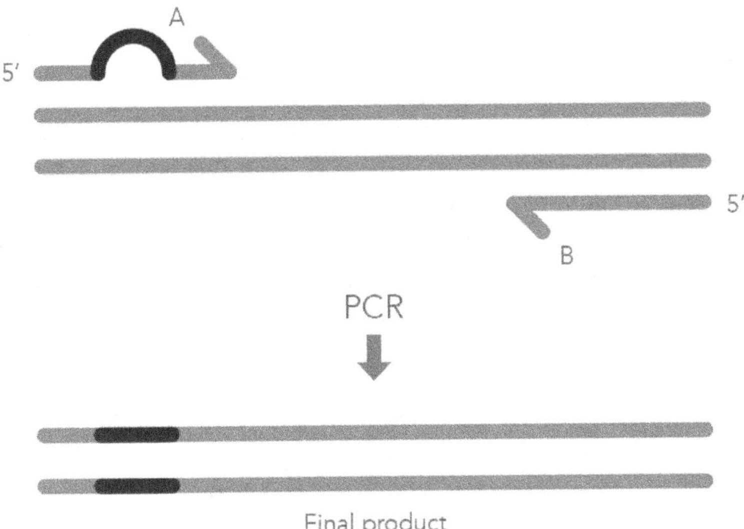

Figure 13.5: PCR for Deletion.

Primer Extension Mutagenesis

Three different amplification procedures and four oligonucleotide primers are needed for the primer extension approach. Two flanking primers amplify the mutant fragment and make it easier to clone the PCR fragment into a suitable vector. Two complementary mutagenic primers cause the mutation into the desired DNA sequence. With this method, several mutations can be produced, including insertions, deletions, and single base-pair modifications. First, two distinct parallel PCR reactions are set up. The "sense" mutant primer and an "anti-sense" flanking primer are both present in one reaction that is 32 to the mutation site. The sense flanking primer 52 to the mutation site and the anti-sense mutant primer are both present in the other reaction. Both of the amplified segments had alterations at either the 52 or 32 termini. The two purified fragments from the first round of PCR are utilised as templates for amplification in the second cycle utilising just the flanking primers. After amplification, thetarget DNA segment, which is then cloned into the proper vectors for DNA sequencing and subsequent functional tests, contains the mutation.

For Insertion

The inserted sequence is present in primers B and C and is denoted by the blue line. Two PCR reactions using the primer pairs A/B and C/D are used in the first round. For the second cycle of PCR, the two produced PCR products are combined using primer pair A/D. The strands can hybridise in the overlapping areas of the two first-round PCR products, and the second round of PCR produces the final, full-length product with the desired insertion (Figure 13.6).

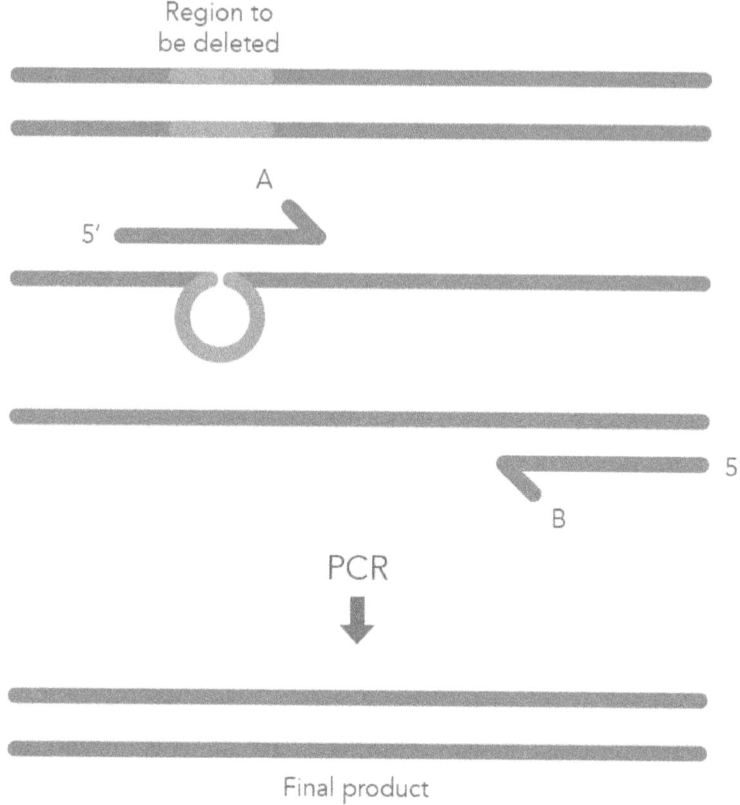

Figure 13.6: Primer Extension for Insertion.

For Deletions

Primers B and C include sequences from both sides of the deletion (shown by dark grey and light grey tails) and are situated on either side of the sequence that will be deleted (orange). After the initial round of PCR, these sequences will enable them to overlap with the other fragment. Primer pairs A/B and C/D are used in the first round of PCR. For the second cycle of PCR, the primer pair A/D is combined with the two resultant PCR products. The strands can hybridise in the overlapping areas of these two first-round PCR products, and the second round of PCR produces the final, full-length product with the desired region removed (Figure 13.7).

Inverse PCR Mutagenesis

Using primers orientated in the other direction, inverse PCR allows for the amplification of an area of unknown sequence. Plasmid mutation is accomplished via inverse PCR. In this technique, the whole plasmid is amplified using two back-to-back primers, and the linear result is then ligated back into the circular form. By changing the primer sequences to include the desired mutation, the primer binding areas may be modified.

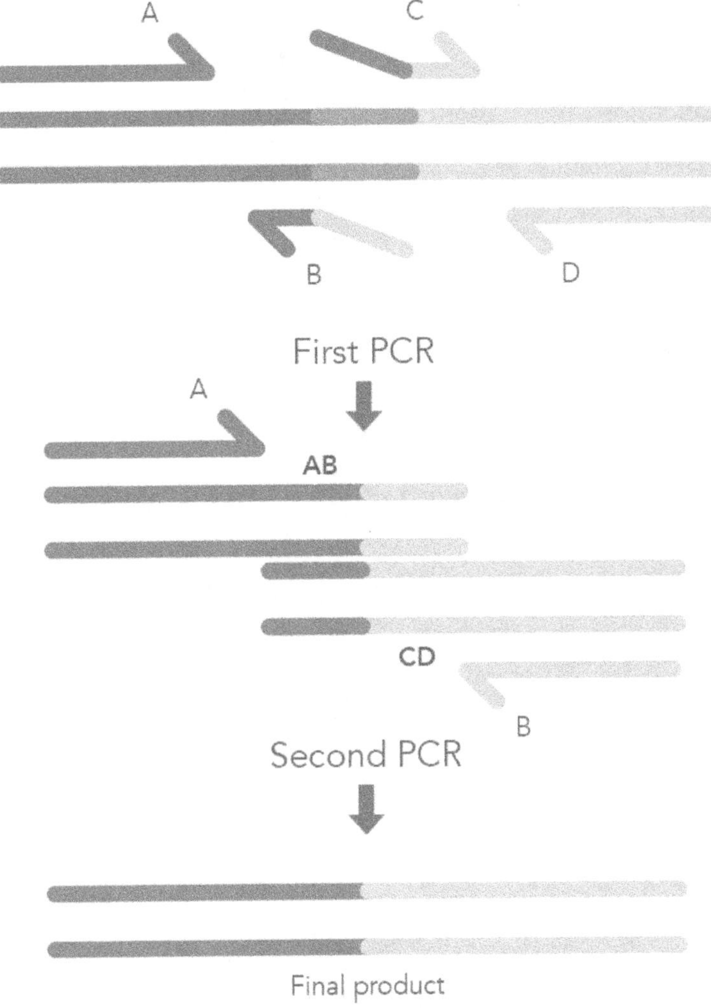

Figure 13.7: Primer Extension for Deletion.

For Deletion

Primers that hybridise to the areas on either side of the deleted area are used in this technique. The primers in this instance had 5′ phosphorylated ends, enabling ligation of the two ends after amplification. High-fidelity DNA polymerase PCR generates a linearized fragment without the deleted region but with blunt ends. Intramolecular ligation is used to circularise this fragment, and the resultant plasmid is subsequently transformed into host bacteria for proliferation (Figure 13.8).

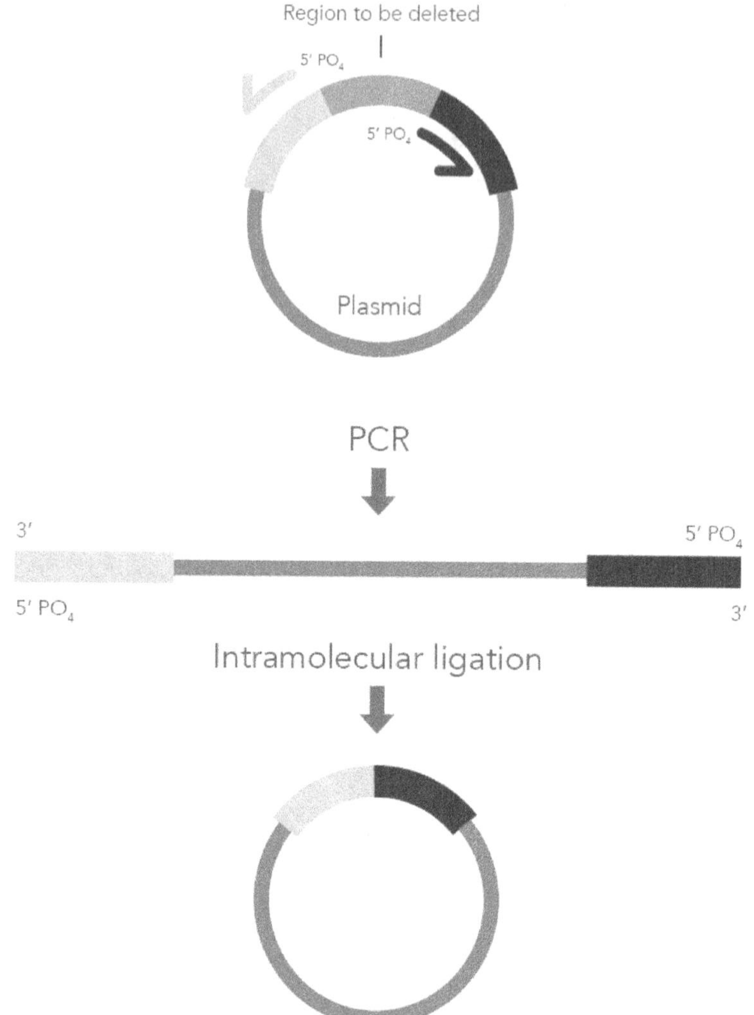

Figure 13.8: Inverse PCR for Deletion.

For Insertion

On either side of the region (depicted by the black, dotted line) where the new sequence will be introduced, the primers are arranged back-to-back. The extra sequence that will be inserted is included in one primer and is denoted by the blue line. Both primers include 5′ phosphorylated ends to make ligation easier after amplification possible. A linearized fragment containing the new sequence is produced by PCR. Following intramolecular ligation, the plasmid is circularised and converted into host bacteria for proliferation (Figure 13.9).

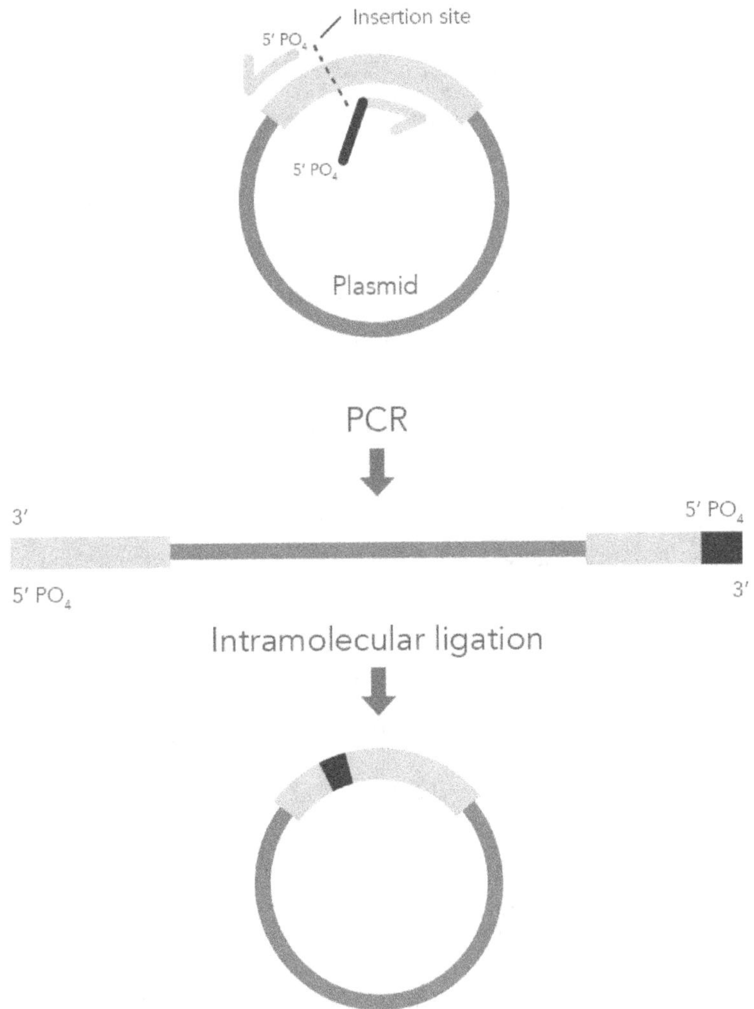

Figure 13.9: Inverse PCR for Insertion.

For Substitution

The desired mutation is present in one of the two primers, as seen by the blue bubble. In this instance, both primers had 5′ phosphorylated ends to enable ligation of the two ends after amplification. The complete circular plasmid is amplified using PCR to produce a linear template that contains the replaced region. Intramolecular ligation is used to circularise this fragment, and the resultant plasmid is subsequently transformed into host bacteria for proliferation (Figure 13.10).

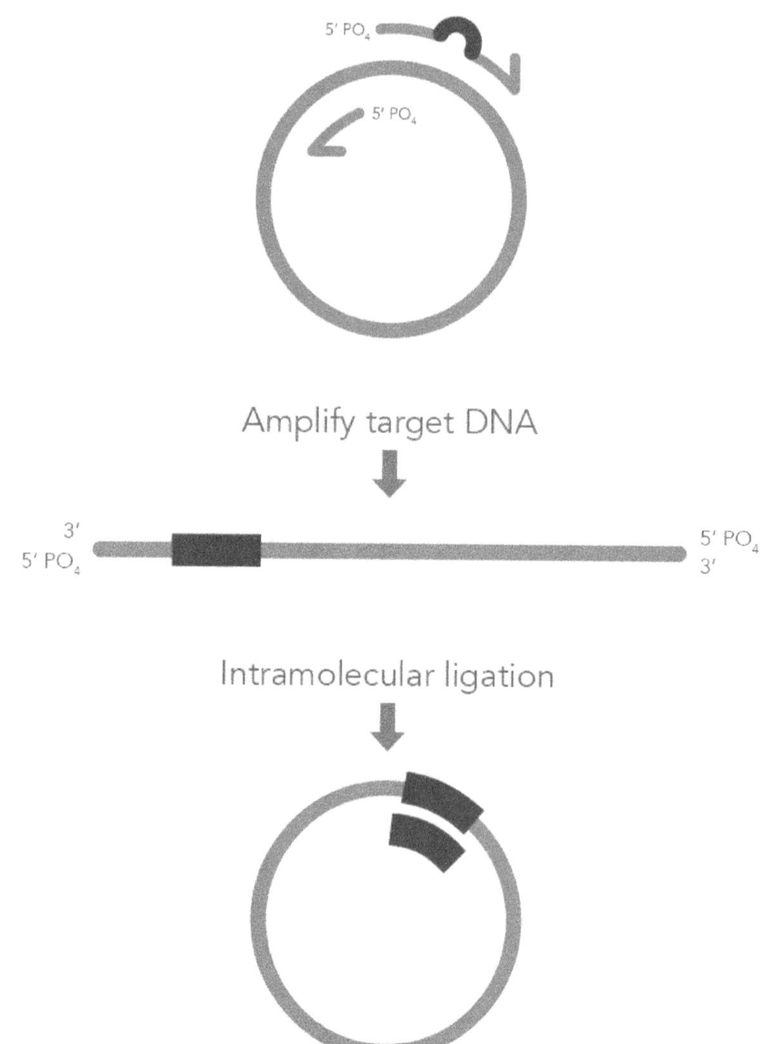

Figure 13.10: Inverse PCR for Substitution.

Nucleases Based Site Directed Mutagenesis

A potent molecular biology method called nucleases-based site-directed mutagenesis is used to induce certain mutations into a DNA sequence at specified sites. Nucleases, specialised enzymes, are used in this procedure to specifically cleave DNA at a chosen place, after which the cellular repair system incorporates the required genetic alterations. The CRISPR-Cas9 system, zinc finger nucleases (ZFNs), and transcription activator-like effector nucleases (TALENs) are the nucleases that are most frequently utilised for this purpose.

Important genomic tools called **transcription activator-like effectors (TALEs)** have programmable DNA binding patterns for locus-specific alterations. TALEs are naturally occurring transcription factors that were discovered in the Xanthomonas plant pathogen. A tract of almost identical repeating units (33–35 amino acid residues) and a partial (or half) repeat unit at the end make up each TALE's DNA-recognition TALE domain. The binding selectivity of each unit towards a DNA nucleotide is completely determined by the two repeat-variable di-residues (RVDs) in a very predictable manner. The RVDs NI, NN for adenine, HD for cytosine, NK, NN, NH for guanine, and NG for thymine are frequently employed. The TALE domain may be built to target practically any DNA sequence in the genome due to its 1:1 RVD to nucleotide modularity. In order to induce DSBs into the target DNA sequence, it also possesses a FokI Nuclease Domain. FokI's ability to cleave needs dimerization, which is the joining of two molecules together. By organising certain amino acids in a way that matches the sequence of the DNA bases, the TALE domains are designed to recognise the target DNA sequence. One base pair is recognised by each TALE domain. A DNA-binding protein that can recognise a longer DNA sequence can be made by joining many TALE domains (Figure 13.11a).

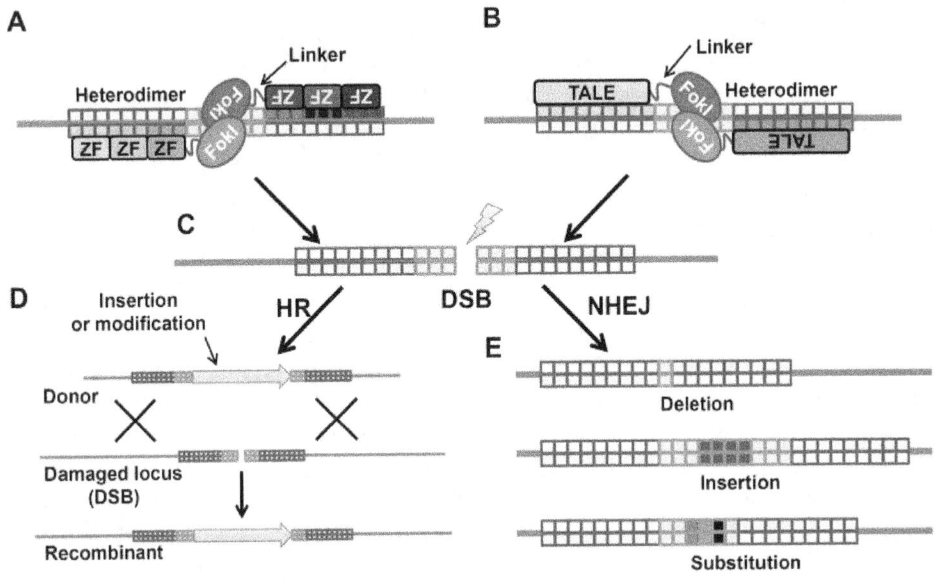

Figure 13.11a: Transcription Activator-like Effector Nucleases (TALEN) b) Zinc Finger Nucleases (ZFN).

Zinc Finger Nucleases (ZFNs) are a class of designed DNA-cleaving proteins that have been widely applied to site-directed mutagenesis and genome editing. ZFNs are made to specifically target certain DNA sequences and cause double strand breaks (DSBs) there. ZFNs are made up of two basic parts: 1) Zinc Finger Domains, which are tiny protein motifs that may be designed to recognise a specific DNA triplet or base pair. Normally, each zinc finger attaches to three DNA bases. It is feasible to design a unique DNA-binding protein that can recognise a longer DNA sequence by fusing many zinc finger domains, and 2) FokI Nuclease Domain: An endonuclease, or DNA-cutting enzyme, is FokI. To produce DSBs in the target DNA, it is employed in ZFNs. FokI is not very precise in targeting a specific DNA sequence, though. The zinc finger domains that direct the FokI domain to the chosen target site help to achieve specificity. To build a useful DNA-cleaving complex, the FokI nuclease domain must dimerize (bind together). By using this characteristic, ZFNs may make sure that two ZFNs attach to target sites on DNA that are near to one another, bringing the FokI domains together. The specificity of ZFNs is increased by this dimerization-dependent cleavage since both ZFNs must attach appropriately for the FokI domains to break the DNA (Figure 13.11b.)

CRISPR (Clustered Regularly Interspaced Short Palindromic Repeats)

It has become a fast programmable molecular tool for selecting certain DNA sequences, offering a technique for marker-less selection while undertaking site-directed mutagenesis. A guide RNA sequence (gRNA), which leads the Cas nuclease to its target, and a CRISPR-associated (Cas) nuclease, which binds and breaks DNA, are the two essential components of the system. In bacteria and archaea, adaptive immunity is based on the CRISPR system. It makes use of Cas nucleases, enzymes that can attach to DNA and cause double-stranded breaks (DSBs). A protospacer, a segment of viral DNA, gets cut off by the Cas nuclease when a virus infects a bacterium. This fragment, along with others from earlier infected viruses, is kept in the bacterial genome as an immunological memory. The arrangement of spacers and palindromic repeats that results in the placement of these viral spacer pieces between repetitive sequences is what gives CRISPR its name. The bacterium may recognise the same virus if it re-infects and use Cas9 to eliminate it. A CRISPR RNA (crRNA) and a trans-activating CRISPR RNA (tracrRNA) are required for Cas9 activation. The tracrRNA functions as a scaffold, and the crRNA is complementary to the viral spacer that was conserved during the first infection; together, these two RNAs make up a complex known as a guide RNA (gRNA). The gRNA serves as the hand directing the scissors of Cas9 when cutting. Using the protospacer adjacent motif (PAM), a brief sequence located downstream of the target site, the Cas9 functions as a search tool before cutting. When Cas9 detects PAM, it scans the area upstream; if it finds the target that the gRNA has supplied, it will produce a double-stranded break (DSB). Because viruses lack their own DNA repair systems, DSBs render the virus ineffective.

The molecular mechanism that made it possible for the CRISPR-Cas9 system to cut DNA naturally was uncovered by Drs. Doudna and Charpentier. Any desired site in the genome of an organism might be sliced once Cas9 is given a distinct guide RNA sequence. A simple and affordable approach of genetic modification, the chimeric single guide RNA (sgRNA) was created by exploiting the native gRNA complex. Cas9 may be used to make cuts at a variety of target locations in the DNA of any organism by simply providing a new guide RNA, which can be produced relatively easily (depending on the existence of the proper PAM sequence). By making double-stranded breaks in the DNA and then using cellular DNA repair mechanisms, CRISPR-Cas9 gene editing achieves its desired results. Despite the fact that there are other DNA repair mechanisms, non-homologous end joining (NHEJ) and homology-directed repair (HDR) are the two most important ones for gene editing. Genes can be rendered inactive by NHEJ, but new genes or genetic material can be inserted through HDR. It was created as a tool for editing genomes as soon as the molecular basis for its capacity to cleave DNA was identified. CRISPR is significant because it enables researchers to change practically any organism's genetic coding. Compared to earlier gene editing methods, it is easier, less expensive, and more accurate (Figure 13.12).

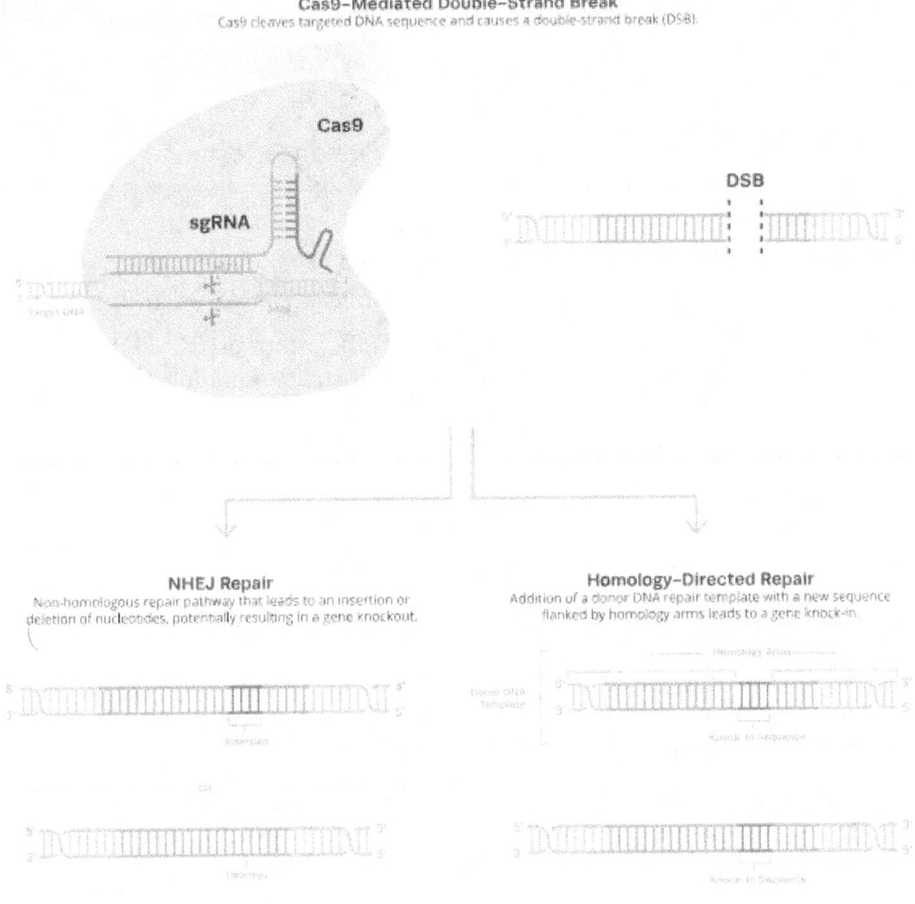

Figure 13.12: CRISPR - Cas.

Conclusion

Site-directed mutagenesis has emerged as a transformative tool in the realm of crop improvement, offering precise and controlled genetic modifications to achieve desired traits. This technique has demonstrated its potential to address key challenges in agriculture, ranging from enhancing nutrient content to fortify staple foods, bolstering disease resistance to safeguard crop yields, and optimizing agronomic traits to increase overall productivity. By allowing researchers to make targeted changes at the molecular level, site-directed mutagenesis has paved the way for a new era of sustainable and resilient agriculture. Through the strategic manipulation of specific genes, scientists can fine-tune crop characteristics while minimizing unintended side effects. This precision not only expedites the development of novel plant varieties with desired attributes but also reduces the reliance on traditional breeding methods that often require years of trial and error. Moreover, site-directed mutagenesis holds the potential to foster crop varieties that can thrive under changing climatic conditions, contributing to global food security.

REFERENCES

Braman J, Papworth C, Greener A. (1996). Site-directed mutagenesis using double-stranded plasmid DNA templates. *In vitro mutagenesis protocols*: Springer; 31–44

Gaj T, Gersbach CA, Barbas III CF. (2013). ZFN, TALEN, and CRISPR/Cas-based methods for genome engineering. *Trends in biotechnology*, 31(7): 397–405.

Gupta M, Gerard M, Padmaja SS, Sastry RK. (2020). Trends of CRISPR technology development and deployment into Agricultural Production-Consumption Systems. *World Patent Information*, 60: 101944.

Ho SN, Hunt HD, Horton RM, Pullen JK, Pease LR. (1989). Site-directed mutagenesis by overlap extension using the polymerase chain reaction. *Gene*, 7(1): 51-59. doi: 10.1016/0378-1119(89)90358-2

Hutchison CA III, Phillips S, Edgell MH, *et al.* (1978). Mutagenesis at a specific position in a DNA sequence. Journal of Biological Chemistry 253 (18): 6551–6560.

Li T, Liu B, Chen CY, Yang B. (2016). TALEN-mediated homologous recombination produces site-directed DNA base change and herbicide-resistant rice. Journal of Genetics and Genomics, 43(5): 297–305

Ochman H, Gerber AS, Hartl DL. (1988) Genetic applications of an inverse polymerase chain reaction. *Genetics*, 120(3): 621-623. doi: 10.1093/genetics/120.3.621

Reikofski J, Tao BY. (1992). Polymerase chain reaction (PCR) techniques for site-directed mutagenesis. *Biotechnology Adv*,10(4): 535-547. doi: 10.1016/0734-9750(92)91451-j

Saboulard D, Dugas V, Jaber M, Broutin J, Souteyrand E, Sylvestre J, *et al.* (2006). High-throughput site-directed mutagenesis using oligonucleotides synthesized on DNA chips. Biotechniques. 2006;39(3): 363–8

Steffens DL and Williams JG (2007) Efficient site-directed saturation mutagenesis using degenerate oligonucleotides. *Journal of Biomolecular Techniques*,18 (3): 147–149.

Zoller MJ. (1991), New molecular biology methods for protein engineering. *Curr Opin Biotechnol.*, 2(4): 526-531. doi: 10.1016/0958-1669(91)90076-h.

Chapter 14

Haploid Production: Techniques and Mutagenesis

Revanna Swamy K.M.[1], Ellandula Anvesh[1], Sree Vathsa Sagar U.S.[1] and Devappa[2]

[1]*Ph.D. Scholar, Department of Genetics and Plant Breeding, TNAU, Coimbatore*
[2]*Ph.D. Scholar, Soil and Water Conservation Engineering, AEC and RI, TNAU, Coimbatore*

INTRODUCTION

Agriculture produces biomass, which is the primary source of human food, animal feed, and useful commodities like wood for building, industrial starches, and fiber for clothing. Our cultivation techniques produced larger yields at the time of Green Revolution in the second part of the previous century. Currently, it is essential that the yields improve substantially in the future due to projections of world population growth, varying environmental conditions, and the agriculture sector's claim on arable land. The production must rise by twofold to ensure food security in 2050 [1].

Aside from the unattractive option of enlarging the farming land, there are two ways to boost production. The gap between achievable and actual yields may be reduced, first, by improving management strategies, such as through precision farming or increasing water availability and nutrients to marginal lands. By improving genetically modified crops through plant breeding, which increases potential yield, is the second and widely regarded as most sustainable technique to enhance outputs [2]. Plant breeding allows for the incorporation of novel features, which is crucial for establishing yield stability in the face of our changing environmental conditions, and may be even more significant. Therefore, the emphasis has shifted to breeding for agricultural traits including fertilizer and water usage efficiency, resistance to salt stress, heat stress, and the capacity to increased and steady yields under stressful conditions. In order to accomplish a doubling in the rate of yield improvement, modern plant breeding must not only include new features into our crops but also speed up the genetic progress of its breeding programs [3].

It cannot be stressed how important and widely applicable haploid and doubled haploid (DH) approaches are to basic and applied research, and how their potential impact on breeding programs ability to accelerate genetic gain cannot be understated and has been the focus of various studies [4]. A variety of crops breeding projects are being accelerated through the application of DH methods, most notably barley (*Hordeum vulgare L.*) and maize (*Zea mays L.*) [5]. This chapter attempts to demonstrate the significance of devoting time and resources to enhancing the effectiveness and efficiency of haploid and DH crop breeding approaches.

Importance of Haploids and Doubled Haploids

Plants with a single set of chromosomes are referred to as haploids (n), while 100 per cent homozygous individuals derived from chromosome doubling of haploids (n) are referred to as doubled haploids (2n). Haploids may be grouped into two broad categories: (a) Monoploids (monohaploids), which possess half the number of chromosomes from a diploid species, *e.g.* maize, barley; and (b) polyhaploids, which possess half the number of chromosomes (gametophytic set) from a polyploidy species, *e.g.* potato, wheat. Here, the general term haploid is applied to any plant originating from a sporophyte (2n) and containing (n) number of chromosomes.

Haploid and DH techniques have found wider application, mostly in the field of crop improvement, when it comes to crops where the protocols are adequately effective. Since DHs reduce the requirement for the 5-7 generations of selfing that are typically needed to establish inbred lines, significant time savings in cultivar development have been gained [6]. The effectiveness of backcross breeding has greatly boosted when DH induction is combined with marker assisted selection (MAS). The character to be introgressed and genotypes with the largest percentage of the elite genome can be quickly selected by using DH induction on one of the early backcross generations. To take advantage of hybrid vigor (heterosis), DHs have been introduced directly as cultivars in barley, rice, rapeseed, wheat, and other crops. They have also been utilized as parents in F1 hybrids of maize and vegetables. As a result of their reduced size, haploid plants have their own market value in ornamental industry [7].

Additionally, DH plant populations have been crucial for QTL identification, particularly in cereals, as their immortality allows for the collection of reliable phenotyping data over a long period of time and in various environments. It has shown successful to produce mapping populations in outcrossers, which experience inbreeding depression. For example, haploids or DHs have been utilized in genome sequencing studies of the peach citrus, coffee, apple, and pear to simplify assembly. Microspores of barley, wheat, rapeseed, and tobacco, are used to fix mutations and transgenes in a single step through successive DH induction [8]. For instance, research on microspore mutations in Brassicas has made it possible to alter the plants fatty acid content, disease resistance, and tolerance to the cold. [9]. In vitro microspore culture technologies have also made it feasible to investigate embryogenesis, early cell fate determination, embryogenesis, and totipotency in great detail [10]. DH and haploid methods can be widely used in different crops in future.

Methods in Obtaining Haploids and Double Haploids

The preferred technique for creating haploid or DH plants varies depending on the species and is influenced by protocol accessibility as well as cost- and yield-effectiveness. Although, haploids could be produced following delayed pollination, irradiation of pollen, temperature shocks and other different methods, these techniques can be grouped into two main categories, in vitro and in vivo methods, depending on whether they include exclusively in vitro procedures, or there are stages of in vivo development of haploid/DH individuals.

In vitro Techniques for Haploid Production

a) Androgenesis

Haploid production from male gamete is most popular and used method. It was reported that haploid and doubled haploid was successfully practiced more than 200 species most of them are belongs to annuals [11]. Methods based on androgenesis exploit the possibility of switching the developmental fate of pollen precursors towards embryogenesis. The most used pollen precursors are, by far, microspores/young pollen grains. Microspore embryogenesis (also known as pollen embryogenesis) is by far the most used and efficient way to produce DHs in vitro. This experimental pathway was first discovered by Guha and Maheswari in 1964 [12], while working with in vitro cultured anthers of Datura innoxia. The success rate is more in solaneace, graminea family compare to leguminasae and perennial woody plants.

In microspore embryogenesis, androgenic haploids/DHs are typically produced upon deviation of microspores towards embryogenesis or callus formation. However, not all species can be induced by isolation and culture of microspores at the same stage. Indeed, for few species it has been shown that male-derived haploid and DH plants can be regenerated from meiocyte-derived callus. Meiocyte-derived callogenesis has been documented in *Arabidopsis thaliana, Vitis vinifera Digitalis purpurea* and *Solanum Lycopersicum* [13]. For some species, mostly cereals, the early stages of microspore development have been described as the only stage where embryogenesis can be induced, However, the majority of works point to the fact that the inducible developmental window revolves around the first pollen mitosis which means that vacuolated microspores but also young, just divided pollen grains would also be inducible. The identification and isolation of these stages is essential for the success of this process, since in the vast majority of species, they are the most sensitive to the induction treatments. Beyond these stages, induction has only unusually been reported [14,15].

To be induced to embryogenesis, microspores/pollens must be stressed. The need for application of physicochemical stress treatments seems common to all inducible species. The variety of responses, depending principally on the genotype but also on the developmental stage of the microspore/pollen, makes that each species has its own specific inductive treatments to trigger the developmental switch. Some of these stresses (heat, cold or starvation) are common to many species, whereas others need more specific stressors or combinations of them as a rule of thumb, the more recalcitrant a species is, and the more combined and more intense stresses are needed [16].

The process of microspore embryogenesis and the different factors, mentioned above, that influence its success, may be implemented in practice using two in vitro approaches: anther culture and microspore culture.

b) Anther Culture

The method used by anther culture is relatively quick, easy, and effective. Early flower buds with immature anthers in which the pollen growth is at the appropriate stage, microspores are contained within the anther sac, are surface sterilized and rinsed with sterile water. The calyx from the flower buds is removed by flamed forceps. The stamens are cut from the corolla, taken out, and put in a clean petri dish. To determine the pollen development stage, one of the anthers is crushed with acetocarmine. Each anther is gently detached from the filament if it is determined to be the right stage, and the intact, unharmed anthers are inoculated horizontally on nutritive media. Anthers that have been wounded may be discarded because injury frequently causes the anther wall tissue to callus. It may be required to utilize a stereo microscope when working with plants that have tiny blooms, such as Brassica and Trifolium, in order to dissect the anthers. Cereal spikes are collected and surface sterilized during the immature stage of microspore formation. The flag leaf covers the inflorescence of the majority of cereals in this state. Anthers can be plated in petri dishes on solid agar material. In a 6 cm petri dish, 10–20 anthers are typically plated. Anthers can be grown on a liquid medium, and 10 ml of liquid medium can culture 50 anthers. The pressure of the expanding pollen callus or pollen plants causes the wall tissue of responding anthers to gradually turn brown and burst open in 3–8 weeks. The individual plantlets or shoots originating from the callus are divided once they have grown to a height of around 3-5 cm, and then they are moved to a medium that would encourage continued development. The rooted plants are moved to the pots' sterile soil mixture. [17].

However, anther cultures are not devoid of limitations. Perhaps, the main limitation comes from the fact that microspores are cultured together with anther walls. Anther walls (the tapetal layer mostly) may secrete molecules that may protect microspores or promote their growth, but it may also secrete inhibitory or even toxic compounds, as is the case of necrosing anther tissues. In any case, this secretory effect is uncontrollable in essence, and makes difficult a strict control of culture conditions. Moreover, when exposed to growth regulators, these walls are able to proliferate in vitro, producing calli. Indeed, some parts of the anther, such as the filament insertion, are especially prone to form calli when in vitro cultured. Therefore, we cannot rule out the chances of occurrence of somatic embryos (very rare but possible) and calli (much more frequent) from anther walls. This implies that for every single plant confirmed as diploid (2C DNA content) by flow cytometry, we should check its origin. For this, the most reliable approach is the use of molecular markers previously confirmed as heterozygous in the donor plants used. However, for very well-known cultivars, where repeated analyses have shown that no somatic embryos are produced, this step may be skipped simply by discarding all calli produced and using only embryos [18].

c) Pollen (Microspore) Culture

Through the in vitro growth of male gametophytic cells, such as microspores or immature pollen, haploid plants can be created. Anthers are often harvested from sterilized flower buds and placed in a small beaker with basal medium (for example, 50 Nicotiana anthers in 10 ml media) as part of a standard technique for microspore culture. The microspores are then squeezed out of the anthers by pressing them against the side of beaker with a glass rod. Anther tissue debris are removed by filtering the suspension through a nylon sieve having a pore diameter which is slightly wider than the diameter of pollen (*e.g.*40 u for Nicotiana, 100 u for maize etc.). Since smaller microspores do not regenerate, it is possible to concentrate larger, good, and viable microspores by filtering the microspore suspension through nylon sieves. This pollen suspension is then centrifuged at low speed ca. 150x g for 5 min. The supernatant of pollen is resuspended in fresh media and it is washed at least twice. The microspores obtained are then mixed with an appropriate culture medium at a density of 103 to 104 microspores/ml. The last step is to pipette the suspension into smaller petri plates. The liquid layer in the dish should be as thin as possible to ensure optimal aeration. After that, each dish is incubated and parafilm-sealed to prevent dehydration. Transferring the embryos or calli that are formed by the responding microspores to an appropriate medium will enable plant formation. [19].

Haploid induction has proven to be successfully accomplished by anther culture. However, one major drawback is that, especially in dicots, plants can develop from a variety of anther components in addition to pollen, producing a population of plants with different ploidy levels. By culturing isolated microspores, this problem can be solved, and the following benefits result: 1. The anther wall and other related tissues' uncontrollable effects are reduced, allowing for better regulation of the numerous elements affecting androgenesis. 2. Beginning with a single cell, the development of androgenesis may be seen. 3. Microspores can be equally exposed to chemical or physical mutagens, making them perfect for absorption, transformation, and mutagenic experiments. 4. It was also possible to produce plants with higher yields per anther.

d) **Gynogenesis**

Gynogenesis would exploit the ability of egg cells to develop in the embryo sac as a haploid zygote without fertilization. This alternative to the normal development of the megagametophyte was first described in vitro in 1976 by San *et al.*, [20]. It would therefore be a form of female haploid parthenogenesis (from the Greek words parthenos, meaning "virgin" and genesis, meaning "origin". In some species, the gynogenic embryo is believed to originate from antipodal or synergid cells, but in the majority of cases the gynogenic embryo is derived from the egg cell. Gynogenic embryos are mostly haploid, which implies that in order to obtain the desired double haploid, the application of additional treatments for chromosome duplication should be considered in nearly all cases. As in the case of androgenesis, colchicine is the most effective and therefore most frequently employed antimitotic which will be discussed later in this chapter.

From a methodological point of view, this technique is implemented by in vitro culture of ovules, ovaries, or even full immature flowers, not yet open and therefore unpollinated, until the embryo sac matures and the gynogenic embryo develops. In some species, mere in vitro culture seems not enough, and an "extra" factor must be applied to trigger the process. Examples of this factor include pollination before ovary excision and in vitro culture, in vitro pollination with mentor pollen from other species, with pollen irradiated with gamma or X rays to inactivate its fertilization capability or with triploid pollen, still able to germinate and stimulate the egg cell, but not to fertilize. In vitro pollination can be done at the apical part of the stigma of entire pistils isolated from the flower and cultured in vitro. After several months, gynogenic embryos will be visible. Alternatively, pollen may be applied by placental pollination, which implies isolating the ovules from the ovary, but maintaining a fragment of the placenta to help the viability of the egg cell. In many cases of gynogenesis in cucurbits and fruit trees, treated pollen is applied directly in situ, on the emasculated flower in the plant. Then, seeds or haploid embryos are rescued and in vitro cultured [21].

In vivo Methods of Haploid Production

a) Haploidization by Wide-Crosses

Wide crosses consist to force crosses between species spanning wide taxonomic boundaries. It could thus involve inter-generic or inter-specific pollinations, and often concern crosses between a cultivated crop and a wild-relative species. Wide crosses are also named wide hybridization, and imply to overcome pre-fertilization and post-fertilization barriers. They have been reported more frequently in monocotyledonous species as compared to dicotyledonous. Two different outcomes, both useful for breeders, need to be distinguished from these wide crosses. Firstly, in case of successful hybridization, which could be helped by embryo rescue or other techniques, the production of hybrid embryos is a starting material in order to introduce agronomic traits of interest (disease resistance, stress tolerance. etc.) across species. Second, it might be utilized to create haploid embryos. Due to the unstable nature of hybrid embryos generated by joining two different genetic materials, chromosome elimination from one parent occurs during early embryogenesis in some "wide-crosses". Although the paternal chromosomes are eliminated in most of the cases to give rise to maternal haploid embryos, some rare cases were reported in which the paternal genome remains, being the maternal genome eliminated [22].

A pioneering discovery in haploidization by wide-crosses was the Bulbosum method, which has been well studied and is now widely used in barley breeding. Cross of cultivated barley (*Hordeum vulgare*) using pollen from its wild relative *Hordeum bulbosum* leads to the production of H. vulgare haploid embryos. The success of barley haploid embryo production thanks to wide crosses was then extended to other species, especially using maize pollen which appears to display low-intensity fertilization barriers For example, wheat and triticale haploid embryos are currently obtained in some plant breeding programs by the following crosses: wheat × maize, and triticale × maize respectively. Overall, wide crosses are limited to some crops at breeding scale, but once they are well established methods, they have the advantage to be effective across a wide range of genotypes, as opposed to the in vitro approaches, which are highly genotype-dependent within a given species [23].

b) Haploidization by Intra-Specific Crosses

Pollination with treated pollen, and the use of haploid inducer lines are the two main methods to induce haploid embryos via intra-specific crosses. While methods based on treated pollen usually necessitate haploid embryo rescue due to early seed abortion, the haploid inducer lines present the advantage to be fully in planta because the output is the production of viable seeds containing haploid embryos. Depending on the methods and species considered, haploidization by intra-specific crosses could produce two different kinds of haploid embryos:

maternal haploid embryos with the cytoplasm and nuclear genome from the female parent and paternal haploid embryos, having the cytoplasm of the egg cell (maternal) but the nuclear genome from the male parent. In the latter case, male haploid embryos produced by intra-specific crosses might be additionally useful for other biotechnological purposes beyond conventional production of DH pure lines. Indeed, since mitochondrial defects are behind cytoplasmic male sterility (CMS), CMS is maternally inherited through the cytoplasm. CMS is a valuable tool in hybrid seed production, since it avoids the timeconsuming process of emasculation to prevent self-pollination. This trait is traditionally transferred from one germplasm to another through multiple rounds of backcrossing. Obtaining a nuclear male genome within a "maternal" cytoplasm in just a single cross reduces CMS conversion to just one step, thus accelerating hybrid seed production [24].

Pollinations with treated pollen induce maternal haploid embryos. This method consists in the treatment of pollen, prior to pollination, with physical or chemical agents, irradiation being the most used treatment. Although haploidization via pollen treatments has been reported in more than 15 species, it works ineffectively (low haploid induction rate). Thus, this method is used in plant breeding only when no alternative efficient methods are available, for example in melon and cucumber. For example In 1990 Pandey *et al.*, [25] first reported successful haploid in A. deliciosa, using gamma irradiated pollen for fertilization, the haploid was generated parthenogenetically from induced unfertilized egg. Cobalt-60 used as radiation source which results pollen germination (through the lethaly irradiated pollen) and parthenogenic embryo development. In 2000, Peixe *et al.*, [26] studied the pollen viability in European plum cv. "Stanley" after different dose of gamma irradiation. These irradiated pollens were used to pollinate a European plum (*Prunus domestica* L.), cv. "Rainha Clàudia Verde". It was seen that abnormal embryo and 2n endosperm was developed with 200 Gy irradiated pollen and haploid was induced parthenogenetically but they differentiated up to heart shape embryo.

The haploid inducers are specialized genetic stocks which, when crossed to a diploid (normal) maize plant, result in progeny kernels in an ear with segregation for diploid (2n) kernels and certain fraction of haploid (n) kernels due to anomalous fertilization. Kernels with a haploid embryo have a regular triploid (3n) endosperm, and therefore, these kernels are capable of displaying germination similar to those kernels with a diploid embryo. The foundation for in vivo haploid induction using haploid inducers was laid when Coe (1959) [27] described "a line of maize with high haploid frequency" of 2.3 per cent, designated as "Stock 6." The discovery of Stock 6, revolutionized the DH production in maize: haploids could be induced either in the maternal parent (gynogenesis) or paternal parent (androgenesis). In paternal haploids, the pollinator is the genome donor and the female parent acts as the haploid inducer. In maternal haploids, the genome originates from the seed parent and the pollinator male parent acts as the haploid inducer.

The inducing capacity of inducer lines in maize has been recently tracked to a 4 bp insertion at the end of the coding sequence of a gene named NOT LIKE DAD (NLD)/MATRILINEAL (MTL)/ZmPHOSPHOLIPASE A1 (ZmPLA1). NLD/MTL/ZmPLA1 is specifically expressed in male gametes and encodes a patatin-like phospholipase A localized at the plasma membrane of the male germ unit. The predicted truncated protein is not detectable in inducer lines and loses its plasma membrane anchorage in a heterologous system. How the biochemical function of NLD/MTL/ZmPLA1 relates to its inducing capacity is still unresolved, and either a structural or a signaling function has been hypothesized. Whereas loss of NLD/MTL/ZmPLA1 is sufficient to trigger haploid induction, quantitative trait locus (QTL) analysis demonstrated that additional, currently unknown players take part in the process and influence the efficiency of haploid induction [28].

The salient steps in DH development in maize using haploid inducers are: (1) crossing the source population (usually a hybrid generated using desired lines or F_2 derived by selfing of the hybrid) as female parent with pollen of the haploid inducer; (2) identification of haploid kernels (at the dry seed stage) using the anthocyanin color marker; (3) germination of the haploid seeds; (4) safe application of colchicine or any other effective chromosome doubling agent to the haploid seedlings; (5) proper agronomic management of D0 seedlings and derivation of D1 (DH) seed by self-pollinating D0 plants; and (6) further selection and utilization of DH lines in breeding programs [29].

Haploid plants can be distinguished from diploid plants by characteristics like erect leaves, poor vigor, and sterility. These characteristics can only be observed after sufficient growth of haploid plants. At present in maize haploid kernels identification relies on the presence of a dominant anthocyanin color marker, referred as R1-Navajo (R1-nj), that expresses in the aleurone (the outermost layer of the maize endosperm) as well as in the embryo (scutellum) in the haploid inducer, unlike the source populations, which do not usually have any anothocyanin coloration in the embryo or the endosperm. Thus, R1-nj as a dominant color marker helps in differentiation of monoploid/haploid (n) kernels (with no expression of purple/red colored anthocynanin in the scutellum, but with the typical crown-coloration on the endosperm), from the diploid (2n) kernels (with expression of anthocyanin in both the endosperm and scutellum) [29].

In potato (*Solanum tuberosum* L.), the haploid induction system relies on a cross between a diploid male haploid inducer line (*S. tuberosum* Andigenum group, previously referred to as cultivar S. Phureja) with a tetraploid cultivated potato of interest used as female parent. It thus refers to an interploidy cross (4x potato × 2x haploid inducer line), and the haploid embryos found in some of the viable seeds of this cross are commonly called dihaploids to indicate that they contain two sets of chromosomes (from maternal origin). Such dihaploid plants are not homozygous, but allow breeders to work at the diploid level for simpler genetic analysis/mapping, or for introgression of valuable traits from the wild species. Since wild species are mostly diploid, they could be then crossed with the di-haploid by inter-specific hybridization [27].

Novel Methods of Haploid Production

1) Haploid Produced by Genetic Engineering

A cutting-edge technique called centromere engineering can also help with haploid induction. In this method, a kinetochore protein (Cenh3) was altered using mutagenesis and then introduced through genetic engineering. Functioning of the novel CENH3 protein is impaired in zygotic cells, the CENH3 protein is severely compromised, resulting in the maintenance of chromosome segregation but the elimination of one chromosomal set during successive mitosis. [24].

The CENH3-based haploid inducer lines originate from the manipulation of the centromeric histone protein CENH3 in Arabidopsis thaliana, it was reported that the genome of the parental having the engineered CENH3 is eliminated after the cross with wild-type plants with intact CENH3, creating haploid inducer lines. These CENH3-based haploid inducer lines are thus able to induce either maternal or paternal haploid embryos although they seem more efficient in producing paternal haploid embryos, in Arabidopsis. Although CENH3 is conserved across plant species, efficient translation of this haploid induction method to crops remain to be achieved, since very low haploid induction rates have been observed so far in crops [24].

2) Haploid Production by Genome Editing Tools

Genome editing tools, like zinc finger nucleases, transcription activator-like effector nucleases, and CRISPR system, have been developed and being used in crop plants for various purposes. Among these techniques, the CRISPR/Cas9 system and a few of its variants are useful for modifying a variety of crop properties, including yield, stress resistance, and nutritional quality. Thus, there is a chance to further investigate the haploid production capabilities of the CRISPR/Cas9 system. Recent research suggested that spermatid chromosomal breakage and the deliberate removal of uniparental chromosomes could result in the generation of haploid maize. Wang *et al.* (2019) [30] reported the genome-edited haploids for ZmLG1 and UB2 in the background of B73 by using the CAU5 haploid inducer line carrying the CRISPR/Cas9 cassette for ZmLG1 or UB2, respectively. He called this method "haploid-inducer mediated genome editing," and it has the potential to speed up the breeding of maize by creating genetically pure, genome-edited DH lines with desired features in an elite genetic background.

Chromosome Doubling of Haploids

In contrast to haploid plants, which only have one copy of each chromosome in each cell, diploid plants have two copies of each chromosome. One copy of each chromosome comes from the male parent and the other from the female parent. Meiotic cell divisions cannot take place in the reproductive systems of haploid plants because homologous chromosomal pairs cannot form, which prevents the development of male and female gametophytes and gametic cells. Consequently, haploid plants are typically sterile. Therefore, the goal of chromosomal doubling is to produce a fertile haploid plant from a haploid (n) by creating a doubled haploid plant (2n), which may then be selfed to produce doubled haploid (DH) lines [29].

Mechanism of Chromosomal Doubling

Low frequency spontaneous chromosomal duplication can lead to certain haploid plants becoming fertile. The source population's genotype affects how frequently spontaneous doubling occurs. Haploid plants are given compounds referred to as mitotic inhibitors in order to produce regular and frequent chromosomal doubling. These substances change normal mitosis in such a way that just one cell with two times as many chromosomes is produced as a result. Colchicine, a water-soluble alkaloid derived from Colchicum autumnale bulbs, is a frequently employed chemical. Colchicine causes chromosomal replication to proceed normally during interphase. In the metaphase of mitosis, colchicine binds to tubulins and blocks the production of spindle microtubules. Two sister chromatids in a replicated chromosome are separated during anaphase, but they are unable to travel to the cell's poles and remain in the middle of the cell. A nuclear membrane develops around the unseparated chromosomes during telophase. Consequently, a cell with two times as many chromosomes emerges from mitosis [29].

Antimicrotubular Agents

Colchicine has been widely used in artificial doubling treatments since the first report of this application in 1937. Researchers are looking at alternative compounds, nevertheless, to ensure safe handling due to its hazardous nature and high carcinogenic potential. The following substances also exhibit antimicrotubular activity: the Trifluralin, oryzalin, and pendimethaline are members of the dinitroaniline group. AMP, trifluralin, oryzalin, and pronamide are examples of herbicides that have been employed in in vitro chromosome doubling investigations. These chemical substances bind to tubulin and prevent spindle formation, which disturbs the sister chromatids segregation toward the opposite poles and causes chromosome duplication. (c-mitosis). Nitrous oxide is a safe gas that has been used to double the number of chromosomes in a variety of cereal crops, including maize. After fertilization, plants are given nitrous oxide gas treatment at 300–600 kPa to double the number of chromosomes in the offspring. [31].

Diploidization

The following techniques can be used to diploidize haploids to create homozygous plants.

1. Colchicine

It is an agent that is frequently used to induce chromosomal duplication by acting as a spindle inhibitor. These are some examples of how it can be used:

i) The plantlets undergo treatment with a 0.5 per cent colchicine solution for a duration of 24-48 hours, followed by a thorough washing before being replanted.

ii) Anthers can be placed directly onto a medium containing colchicine for a week. Once the initial division has taken place, they can be transferred to a colchicine-free medium to initiate the androgenesis process. This method is applicable in maize, where separate male and female flowers are produced, and diploidization is a potential issue.

iii) Upon reaching full maturity, a mixture of colchicine and lanolin paste (0.4 per cent) can be administered to the leaf axils of the plants. This application serves to stimulate the development of diploid and fertile branches from the axillary buds. Additionally, the main axis is pruned to promote this process.

iv) The axillary buds undergo a series of colchicine treatments using cotton wool plugs for an extended duration, such as 14 days in the case of potatoes.

v) In cereal plants, when they have reached the 3–4 tiller stage and are in good health, they are collected. After removing the soil from their roots, the plants are pruned to a height of 3 cm below the crown. Using a colchicine solution (consisting of 2.5 g colchicine dissolved in 20 ml of dimethyl sulfoxide and then brought up to a liter with water), glass jars or vials containing the plants are prepared. The crowns of the plants are coated with the colchicine solution. Subsequently, the plant roots are thoroughly rinsed with water and then transplanted into light soil. Before transplanting, the plants are exposed to light at room temperature for 5 hours.

2. Endomitosis

Haploid cells can turn unstable in culture, tending to become diploid via endomitosis. This trait has been utilized to induce diploidy for homozygous plant production. Cultivating stem segments with auxin-cytokinin media encourages callus formation. Callus growth triggers chromosome doubling through endomitosis, yielding diploid cells that mature into plants [31].

Conclusion

In this chapter we have revised the principal approaches currently available to produce haploids and DHs for different purposes, principally focused on the rapid generation of pure lines to simplify breeding programs. These approaches imply the use of methods exclusively based on in vitro culture, in vivo induction and novel methods of haploid development, the choice of the best performing approach will depend on the species used, and to what extent these methods have been developed and adapted to this species. Additionally, this technology speeds up cultivar growth, maximizes genetic gains in breeding programs, and lowers the price of breeding schemes. For the mass production of inbred lines, numerous international seed corporations have embraced DH technology. Thus, the creation of DH plants using novel haploid induction techniques has enormous promise for the management of genetic resources, the improvement of germplasm, and the creation of unique plant populations.

REFERENCES

1. Tilman, D., Balzer, C., Hill, J., Befort, B.L. Global food demand and the sustainable intensification of agriculture. Proc. Natl. Acad. Sci. USA 2011, 108, 20260–20264.

2. Dawson, I.K., Russell, J., Powell, W., Steffenson, B., Thomas, W.T.B., Waugh, R. Barley: A translational model for adaptation to climate change. New Phytol. 2015, 206, 913–931.

3. Tester, M., Langridge, P. Breeding technologies to increase crop production in a changing world. Science 2010, 327, 818–822.

4. Forster, B.P., Heberle-Bors, E., Kasha, K.J., Touraev, A. The resurgence of haploids in higher plants. Trends Plant Sci. 2007, 12, 368–375.

5. Geiger, H.H., Gordillo, G.A. Doubled haploids in hybrid maize breeding. Maydica 2009, 54, 485–499.

6. Dwivedi, S.L., Britt, A.B., Tripathi, L., Sharma, S., Upadhyaya, H.D., Ortiz, R. Haploids: Constraints and opportunities in plant breeding. Biotechnol. Adv. 2015, 33, 812–829.

7. Seguí-Simarro, J.M., Nuez, F. How microspores transform into haploid embryos: Changes associated with embryogenesis induction and microspore-derived embryogenesis. Physiol. Plant. 2008, 134, 1–12.

8. Devaux, P., Kasha, K.J. Overview of barley doubled haploid production. In Advances in Haploid Production in Higher Plants, Touraev, A., Forster, B.P., Jain, S.M., Eds., Springer: Dordrecht, The Netherlands, 2009, pp. 47–64.

9. Ferrie, A.M.R., Möllers, C. Haploids and doubled haploids in Brassica spp. for genetic and genomic research. Plant Cell Tissue Organ Cult. 2010, 104, 375–386.

10. Shen, Y., Pan, G., Lübberstedt, T. Haploid strategies for functional validation of plant genes. Trends Biotechnol. 2015, 33, 611–620.

11. Dunwell JM (1986) Pollen, ovule and embryo culture, as tools in plant breeding. In: Withers LAP, Alderson G (eds) Plant Tissue Culture and Its Agricultural Applications. London: Butterwoorths, pp. 375–404.

12. Guha S, Maheshwari SC (1964) In vitro production of embryos from anthers of Datura. Nature 204: 497

13. Seguí-Simarro JM, Corral-Martínez P, Parra-Vega V, González-García B (2011) Androgenesis in recalcitrant solanaceous crops. Plant Cell Rep 30 (5): 765-778. doi: 10.1007/s00299-010-0984-8.

14. Touraev A, Pfosser M, Heberle-Bors E (2001) The microspore: A haploid multipurpose cell. Adv Bot Res 35: 53-109.

15. Binarova P, Hause G, Cenklova V, Cordewener JHG, van Lookeren-Campagne MM (1997) A short severe heat shock is required to induce embryogenesis in late bicellular pollen of Brassica napus L. Sex Plant Reprod 10 (4): 200-208.

16. Corral-Martínez P, Seguí-Simarro JM (2012) Efficient production of callus-derived doubled haploids through isolated microspore culture in eggplant (Solanum melongena L.). Euphytica 187 (1): 47-61. doi: 10.1007/s10681-012-0715-z

17. Germanà MA (2011) Anther culture for haploid and doubled haploid production. Plant Cell Tissue Organ Cult 104 (3): 283-300. doi: 10.1007/s11240-010-9852-z

18. Corduan G, Spix C (1975) Haploid callus and regeneration of plants from anthers of Digitalis purpurea L. Planta 124 (1): 1-11

19. Touraev A, Pfosser M, Heberle-Bors E (2001) The microspore: A haploid multipurpose cell. Adv Bot Res 35: 53-109

20. San Noeum LH (1976) Haploides d'Hordeum vulgare L. par culture in vitro d'ovaries non fécondés. Ann Amélior Plantes 26: 751-754

21. Kantartzi S, Roupakias D (2009) In vitro gynogenesis in cotton (Gossypium sp.). Plant Cell Tissue Organ Cult 96 (1): 53-57. doi: 10.1007/s11240-008-9459-9.

22. Baum M, Lagudah ES, Appels R (1992) Wide Crosses in Cereals. 43: 117-143

23. Lange W (1971) Crosses between Hordeum vulgare L. and H. bulbosum L. I. Production, morphology and meiosis of hybrids, haploids and dihaploids. Euphytica 20 (1): 14-29. doi: 10.1007/bf00146769

24. Jacquier NMA, Gilles LM, Pyott DE, Martinant J-P, Rogowsky PM, Widiez T (2020) Puzzling out plant reproduction by haploid induction for innovations

in plant breeding. Nature Plants in press: in press. doi: 10.1038/s41477-020-0664-9

25. Pandey KK, Przywara L, Sanders PM (1990) Induced parthenogenesis in kiwifruit (Actinidia deliciosa) through the use of lethally irradiated pollen. Euphytica 51: 1–9.

26. Peixe A, Campos MD, Cavaleiro C, Barroso J, Pais MS (2000) Gamma-irradiated pollen induces the formation of 2n endosperm and abnormal embryo development in European plum (Prunus domestica L., cv. "Rainha Clàudia Verde"). Sci Hort 86: 267–278

27. Coe EH (1959) A Line of Maize with High Haploid Frequency. Amer Nat 93 (873): 381- 382. doi: 10.1086/282098

28. Gilles, L. M., Khaled, A., Laffaire, J. B., Chaignon, S., Gendrot, G., Laplaige, J.,. Widiez, T. (2017). Loss of pollen-specific phospholipase NOT LIKE DAD triggers gynogenesis in maize. EMBO Journal, 36, 707–717. https: //doi. org/10.15252/embj.2017966

29. B.M. Prasanna, Vijay Chaikam and George Mahuku (eds). 2012. Doubled Haploid Technology in Maize Breeding: Theory and Practice. Mexico, D.F.: CIMMYT.

30. Wang, B., Zhu, L., Zhao, B., Zhao, Y., Xie, Y., Zheng, Z., Wang, H. (2019). Development of a Haploid-Inducer mediated genome editing system for accelerating maize breeding. Molecular Plant, 12, 597–602. https: //doi. org/10.1016/j.molp.2019.03.006

31. Maqbool MA, Beshir A, Khokhar ES. Doubled haploids in maize: Development, deployment, and challenges. Crop Science. 2020,60: 2815–2840. https: //doi.org/10.1002/csc2.20261

Chapter 15

Integration of High-Throughput Phenotyping: Phenomics in Modern Plant Breeding

R. Nivedha, T. Nivethitha and Nagendra Naidu*

Ph.D. Scholar, Department of Genetics and Plant Breeding, Tamil Nadu Agricultural University, Coimbatore – 641 003, Tamil Nadu
**e-mail: nivedharakkimuthu@gmail.com*

INTRODUCTION

Plant breeding has been instrumental in advancing agriculture and addressing the challenges of food security, climate change, and sustainable crop production. The genetic improvement of crop plants is made by precise and accurate measurement of phenotype which is the result of interaction between genotype and environment. Plant phenomics is the exploration of plant growth, structure, function, and composition through efficient data collection and analysis. Forward phenomics examines large collections of genetic material for beneficial traits. Conversely, reverse phenomics involves in-depth study of a specific trait, to understand the underlying physiological, biochemical, and genetic mechanisms governing its expression [1]. In addition to phenomics at controlled environment, recent advances in the technologies have paved the way for high-throughput plant phenotyping (HTPP) under natural environment. Field-based crop phenotyping encompasses the systematic assessment and measurement of various plant traits, such as morphometric and physiological parameters, within the actual field environment where crops are cultivated. The integration of high-throughput phenotyping technologies such as remote sensing, imaging, and diverse sensors has ushered in a new era of innovation in plant breeding, enabling breeders to make significant advancements in trait discovery, selection, and variety development.

HTPP Techniques and Technologies

Phenomics at Controlled Environment

The techniques used in High-Throughput Plant Phenotyping in growth chambers or greenhouses can be broadly classified into two categories: "sensor-to-plant" and "plant-to-sensor" based on the positioning of plants during the measurement routine and imaging setup [2].

1. Sensor-to-plant: In this approach, the plants occupy fixed positions within the growth chamber or greenhouse, and imaging devices move around to capture data. In addition, multiple sensors and imaging devices can be mounted to capture the plants.

2. Plant-to-sensor: In this approach, the plants are transported to an imaging station or sensor-equipped area where data is collected. This approach involves moving the plants, one at a time or in batches, to the imaging setup.

Both approaches have their advantages and limitations, and the choice depends on the specific research objectives, available resources, and the scale of phenotyping required.

Phenomics at Field Level

Crop phenotyping in the field environment utilizes ground or aerial platforms mounted with cameras, sensors, and advanced computing. It enables deep phenotyping throughout the crop cycle and canopy level. Equipped with various high-resolution cameras and sensors, these platforms allow comprehensive data collection for better crop analysis. These tools offer high-throughput observations with spatial, spectral, and temporal resolutions that cater to diverse phenotyping demands. Ground and aerial platforms utilized in phenotyping will be provided in the latter section of this chapter briefly.

In HTPPs, a wide range of techniques and technologies are used for data collection and analysis. Some common techniques are tabulated (Table 15.1). These techniques, along with advanced automation, image processing algorithms, and data analytics, enable researchers to conduct high-throughput plant phenotyping, facilitating the analysis of large plant populations efficiently and accurately both at controlled environment and field level.

Table 15.1: Phenotyping Technologies in High Throughput Plant Phenotyping (HTPP's) [2,1]

Imaging Technologies	Sensors/ Camera	Phenotyping Traits
Visible light imaging	RGB Cameras (400-950 nm) (Mimics naked eye perception)	☆ Morphological traits: Canopy height, canopy coverage, number of branches, leaf length, grain and yield related traits *etc.* ☆ Phenological traits ☆ Green area index (GAI) ☆ Weed infestation ☆ Crop growth status, biomass and nutritional status.
Fluorescence imaging	Fluorescence camera (400-950 nm)	☆ Chlorophyll pigments, photosynthesis, lodging, GAI, biomass, grain quality and yield. ☆ Light and nitrogen use efficiency. ☆ Quantification and identification of biotic and abiotic stress.
Near Infra-red imaging (NIR)	NIR Camera (900-1700 nm)	☆ Drought response: root phenotyping and water distribution
Infrared imaging (IR)	Infrared Camera (8,000-14,000 nm)	☆ Biotic and abiotic stress studies ☆ Green area index ☆ Drought response: temperature (leaf and canopy), leaf water content, leaf rolling, leaf wilting and stomatal conductance
Hyper Spectral imaging	Hyperspectral camera (400 – 1700 nm)	☆ Morphological traits ☆ Biochemical analysis ☆ Senescence/stay green traits ☆ Crop growth potential, nutrient and water use efficiency ☆ NDVI ☆ Pest and disease severity.
3D Imaging	Stereo camera	☆ Plant organ morphology: shoot, leaf, canopy structures, biomass, leaf rolling and leaf angle. ☆ Weed infestation ☆ Crop dynamics monitoring
Laser scanning	Instruments for laser scanning	☆ Similar to 3D imaging
Magnetic resonance imaging (MRI)	MRI imagers	☆ Water content and distribution within plants (localization of bound and free water) ☆ Spatial organisation of tissues and organs ☆ Identification of frost damaged and healthy tissues.

Use of Multiple Sensors for Monitoring

In this approach, a combination of sensors, such as visible, thermal, and fluorescence sensors *etc.*, are employed to phenotype plants simultaneously [1]. This multifaceted approach yields more reliable results in identifying phenotypes accurately. In stress studies, utilizing individual imaging sensors can only reveal the symptoms of a wide range of stresses, but the specific stress responsible for a particular symptom can be deciphered only through the use of multiple sensors.

For instance, fluorescence sensors can detect changes in leaf chlorophyll content, indicating a potential stress in the plant. However, this decrease in chlorophyll could be attributed to different factors, such as drought or nitrogen deficiency. To pinpoint the root cause accurately, additional information is required. In this case, using infrared sensors to monitor leaf or canopy temperature becomes crucial. Drought stress causes stomatal closure, leading to an increase in leaf temperature, whereas nitrogen deficiency does not influence temperature. By employing both fluorescence and infrared sensors simultaneously, researchers can determine the actual stress factor responsible for the observed symptom of decreased chlorophyll content.

By integrating diverse sensor data, researchers gain a comprehensive understanding of plant responses to various stresses, enabling more accurate and targeted interventions for crop improvement and management.

1. Phenotyping Platforms

Plant phenotyping platforms can be categorized into two main types: ground-based and aerial-based [3]. Specialized phenotyping platforms at ground and aerial level are equipped with diverse sensors and data communication systems which ensures automatic data acquisition and transmission.

Ground Level Platforms

Ground level phenotyping platforms are broadly classified into three types, which includes phenopoles, phenomobiles and stationary platforms (Figure 15.1).

Phenopoles

It is helpful in recording canopy cover, Green area index, phenology and other morphological traits based on the sensors mounted. Phenopoles are available in fixed and mobile versions.

Fixed phenopoles resemble cost-effective field weather stations. These phenopoles are equipped with RGB sensors that capture images based on sunlight illumination. The recorded images are then regularly updated to a central server every 30 to 60 minutes. Thus, it facilitates easy monitoring of dynamic changes in canopy coverage and GAI. However, a significant limitation

Figure 15.1: Ground Level Phenotyping Platforms [3].

of phenopoles is their limited coverage, as they can only monitor a subsection of the experimental site. To overcome this limitation, multiple cameras are need to be installed across the field *e.g.,* Phenocam.

On the other hand, Mobile phenopoles are designed to be manually carried, which requires manpower for their operation. These mobile setups utilize wireless-fidelity (Wi-fi) technology to control the cameras mounted on them remotely using mobile phones. By capturing images at ground level (1-3 meters), these mobile phenopoles can provide high-resolution images.

Phenomobiles

Phenomobiles are available in various versions including, automatic, human and tractor propelled platforms.

The automated version of phenomobiles is equipped with a power supply unit and integrates various sensors such as RGB, hyperspectral, and multispectral *etc,.* along with GPS to capture images with precise navigation coordinates. The navigation pattern of these phenomobiles is controlled using the navigation control system. To ensure data quality, the data acquisition system regulates the brightness of the flashlight and minimizes background noise during image capture. Prominent example of these automatic phenomobiles is Ladybird at the University of Sydney, Australia.

In contrast, human-propelled phenomobiles are more budget-friendly but require manual movement throughout the field. Their data acquisition system is controlled externally using a laptop. However, the reliability of data quality and overall efficiency of these phenomobiles may not be as consistent as their automated counterparts.

Tractor-propelled phenomobiles share similarities with automatic systems, including the integration of image-based sensors and GPS. However, the sensors mounted are non-imaged based sensors (*e.g.,* green seeker and crop circle). These phenomobiles draw power from the tractor to operate the sensors, GPS and data acquisition system.

Stationary Platforms

This fixed phenotyping platform is a fully automated system mounted on trackways, covering an extensive area of about one hectare. They are equipped with a range of sensors, including RGB, multi/hyperspectral, thermal and laser sensors. These sensors work in synergy to enable simultaneous phenotyping of morphometric traits, physiological traits, crop growth and development and stress-related responses.

Aerial Level Platforms

Aerial level phenotyping platforms are broadly categorized into three main types: MAVs (manned aerial vehicles), UAVs (unmanned aerial vehicles), and micro/nano satellites (Figure 15.2).

Figure 15.2: Aerial Level Phenotyping Platforms [3].

MAVs and UAVs share some similarities, they also have distinct characteristics and limitations. Unlike UAVs, MAVs require human interventions during their operations. Despite this requirement, MAVs offer the advantage of covering larger areas in a relatively shorter time, albeit at the cost of lower image resolution when compared to UAVs. However, what sets MAVs apart is their greater payload capability, enabling them to carry more sensors or equipment for data collection.

On the other hand, satellite-based imaging also suffers from limitations, with low image resolution being one of the primary issues. Moreover, the quality of satellite imagery heavily depends on external weather conditions, which can impact the effectiveness of the data collected. Due to these limitations associated with both MAVs and satellites, UAVs have emerged as the most popular and favored platforms for aerial phenotyping.

Unmanned Aerial Vehicles (UAVs)

Recent technological advancements, particularly in UAVs, have revolutionized high spatial resolution crop phenotyping at the field level. UAVs offer the capability to capture detailed images with four main groups: parachutes, blimps, fixed-wing systems and multirotor setups [4]. Among these, the fixed-wing and multirotor UAVs are the most widely used.

Fixed-wing UAVs and multirotor UAVs differ in several aspects:

1. **Flight Speed and Efficiency**: Fixed-wing UAVs have high flight speeds and are efficient in covering large areas quickly, while multirotor UAVs are slower in comparison.

2. **Payload Capacity**: Fixed-wing UAVs can carry larger payloads, enabling them to accommodate more sensors and equipment. Multirotor UAVs have smaller payload capacities.

3. **Flight Altitude**: Fixed-wing UAVs can fly at higher altitudes, capturing a broader view of the area, whereas multirotor UAVs operate at lower altitudes.

4. **Endurance Time**: Fixed-wing UAVs have longer endurance times, allowing them to remain in the air for extended periods. Multirotor UAVs have shorter flight times.

5. **Taking Off and Landing Requirements**: Fixed-wing UAVs have more complicated take-off and landing procedures, while multirotor UAVs have simpler requirements.

6. **Image Quality**: Fixed-wing UAVs may produce blurred images due to their high-speed capturing, whereas multirotor UAVs generally provide good image quality.

Presently, UAVs are extensively equipped with multiple sensors, enabling the phenotyping of a wide range of traits, including yield estimation, NDVI

(Normalized Difference Vegetation Index), canopy temperature, nitrogen status, fluorescence, and various morphometric traits. The acquired images are processed using software like Agisoft Photoscan Professional (Version 1.2.2) developed by Russia. Geopositioning of the images is achieved through software that recognizes ground control points (GCPs), ensuring accurate spatial referencing and data analysis in the field of crop phenotyping.

Applications of High Throughput Phenotyping

The Big data generated through imaging and remote-sensing can be interpreted through machine learning for high-throughput phenotyping. Machine learning techniques such as Artificial Neural Network, Convolution Neural Network *etc.* along with digital images, could be used to model and predict genotypes responses to different growth conditions. By this way, more resistant genotypes to stress and nonstress environments can be selected.

Early Season Crop Mapping

In the context of early-season crop classification, advanced remote sensing algorithms such as support vector machines, random forests, decision trees, and neural networks are employed [3]. The integration of Unmanned Aerial Vehicles (UAVs) with optical sensors furnishes ultrahigh spatial resolutions to enhance the precision of early-season crop classification. The Synthetic Aperture Radar (SAR) Satellites can be used for prediction since they are not influenced by atmospheric interferences. Other deep learning techniques including convolutional neural networks and recurrent neural networks are harnessed to effectively map and discern various crop types during the initial stages of the growing season.

Stress Monitoring and Detection

Abiotic Stress Detection

Nitrogen

Nitrogen is an indispensable nutrient involved in diverse physiological functions such as photosynthesis, protein synthesis and enzymatic reactions. Insufficient nitrogen supply results in inhibited growth, reduced leaf development, delayed flowering and decreased agricultural productivity.

The near-infrared and visible spectral reflectance techniques are employed in evaluating different nitrogen levels. An integrated approach employing Unmanned Aerial Vehicle (UAV)-borne multispectral and hyperspectral sensors, coupled with fluorescence sensors and phenomobiles, emerges as a proficient means for recording crop traits under nitrogen stress. This integrated approach facilitates early-stage detection and comprehensive evaluation of nitrogen-related physiological responses in crops.

Water

Water stress induces stomatal closure, leading to elevated canopy temperature as a result of decreased transpiration. Utilizing thermal camera mounted UAVs allow efficient acquisition of crop temperature imagery. In addition to this, the effects of drought can be monitored by many assays *viz.*, stomatal conductance, IR thermography, chlorophyll fluorescence and estimation of tissue water content [5]. Optical sensors facilitate biomass estimation in water-stressed conditions and this facilitates the non-destructive selection of drought-tolerant cultivars.

Lodging

Alteration in crop structure or morphology leads to variations in reflectance and backscattering across distinct wavelengths. Ground-based platforms leverage linear polarization of backscatter to detect crop lodging. Unmanned Aerial Vehicles (UAVs) equipped with thermal cameras capture thermal images, aiding in lodging area detection, particularly in wheat [6]. In rice, improved accuracy in lodging detection is achieved through methodologies such as optical hyperspectral data integration, amalgamation of RGB and thermal imagery.

Biotic Stress Detection

Weed Detection

Weeds contribute to the highest yield loss, surpassing the impact of pests and diseases. The excessive reliance on herbicides has given rise to herbicide-resistant weed species. Precise weed detection is of paramount importance, as evidenced by research achieving 88-94 per cent accuracy in rice fields and up to 91 per cent in sunflower through UAV-based methods [7]. Innovations have led to the creation of real-time tractor systems designed for weed detection and targeted herbicide application.

Diseases and Pest Monitoring

The utilization of high-throughput phenotyping provides a non-invasive approach to effectively identify pests and diseases in the field. Detecting these problems before the appearance of visible symptoms will significantly reduce crop yield losses and optimize the use of pesticides. Thus eliminating the need for proactive fungicide sprays and labor-intensive manual symptom scoring across vast fields.

For instance, in the case of wheat, distinct spectral patterns were observed even at an incipient 4-5 per cent disease incidence, yielding a remarkable 96 per cent accuracy [8]. Similarly, for potato virus Y, near-infrared and short-wave infrared proved more efficacious than visible light [9]. Notably, lab-based disease recognition using neural networks achieved an impressive 99 per cent accuracy, which experienced a reduction to 31 per cent under real-world environmental conditions. [4].

Unmanned Aerial Vehicles (UAVs) offer a promising platform for disease detection, facilitating swift coverage of vast areas. Spectral imaging emerges as a potential method for plant resistance assessment, though considerations such as canopy and leaf traits influence disease detection outcomes. Further research is essential to ascertain if computer vision or spectral alterations can accurately identify early disease symptoms. The fluorescence sensors can be integrated with hyperspectral, multispectral or thermal cameras, enhancing early identification and screening of disease-resistant cultivars. This holistic approach promises an advanced detection against agricultural diseases, ensuring enhanced crop health and productivity.

Crop Development Tracking

Analysis of Root System

The evaluation of root traits in field-grown plants commonly involves excavating and visually assessing root systems. Additionally, camera systems inserted through Plexiglas tubes in the soil indirectly reveal root biomass by measuring changes in soil electrical properties caused by root water uptake. Advanced imaging techniques like X-ray computed tomography, neutron radiography, and magnetic resonance imaging are employed for 2-D and 3-D soil-grown root system analyses [1]. Automated phenotyping platforms often use aeroponic or hydroponic systems for root visualization, extracting traits like length and branching angles. These methods contribute to in-depth root trait assessments and plant growth understanding.

Optimizing Fertilization

Fertilization is given based on soil nutrient analysis of the field and the prescribed dosage to the crop. Employing optical sensors for plant phenotyping to pinpoint nutrient-deficient zones within the field signifies a noteworthy advancement in precision agriculture. This approach ensures precise fertilizer distribution and mitigates excessive fertilizer application [4].

Yield Estimation

The pre-harvest crop yield estimation methodologies enhance agricultural output, facilitates optimal resource allocation and contributes to the entire food ecosystem. Moreover, this approach is instrumental in selecting high yielding lines from extensive germplasm screenings. Sensor-driven high throughput technologies harness a spectrum of insights encompassing plant health, growth stages, weather dynamics and historical yield trends to estimate the crop yield. Advanced machine learning techniques, particularly deep learning algorithms like Convolutional Neural Networks (CNNs) and Long Short-Term Memory (LSTM) neural networks exhibit the potential to elevate the precision of yield forecasting. Few examples of crops and the respective phenotyping tools used for yield detection are listed as follows [4].

Crop	Phenotyping Tool Used
Maize	Field hyperspectral imaging [10]
Wheat	Combined field spectral VIs with meteorological data through hierarchical linear modelling [11]
Soybean	UAV RGB multispectral and thermal images (multimodal data fusion and deep learning methods) [12]
Rice	UAV RGB multispectral imaging (deep CNN) [13]

Conclusion

The accomplishments realized through high-throughput phenotyping are phenomenal. Nevertheless, a spectrum of challenges continues to persist. The realms of data management, standardization, scalability, and the harmonious amalgamation of diverse data streams represent persistent issues demanding attention. Overcoming these challenges requires the collaboration of innovative technological advancements and creative approaches for interpreting and leveraging data.

REFERENCES

1. Singh B D and Singh A K. Marker Assisted Plant Breeding: Principles and Practices. Springer, New Delhi, 2015, 431-450.

2. Zhao C, Zhang Y, Du J, Guo X, Wen W, Gu S, Wang J, Fan J. Crop phenomics: current status and perspectives. Frontiers in Plant Science. 2019. 10: 714.

3. Jin X, Zarco-Tejada PJ, Schmidhalter U, Reynolds MP, Hawkesford MJ, Varshney RK, Yang T, Nie C, Li Z, Ming B, Xiao Y. High-throughput estimation of crop traits: A review of ground and aerial phenotyping platforms. IEEE Geoscience and Remote Sensing Magazine. 2020. 9(1): 200-231.

4. Chawade A, van Ham J, Blomquist H, Bagge O, Alexandersson E, Ortiz R. High-throughput field-phenotyping tools for plant breeding and precision agriculture. Agronomy. 2019. 9(5): 258.

5. Munns R, James RA, Sirault XR, Furbank RT, Jones HG. New phenotyping methods for screening wheat and barley for beneficial responses to water deficit. Journal of Experimental Botany. 2010. 61: 3499–3507.

6. Chapman S C *et al.*, Pheno-copter: A low-altitude, autonomous remote-sensing robotic helicopter for high-throughput field-based phenotyping. Agronomy. 2014; 4 (2): 279–301.

7. Peña J *et al.*, Quantifying Efficacy and Limits of Unmanned Aerial Vehicle (UAV) Technology for Weed Seedling Detection as Affected by Sensor Resolution. Sensors. 2015; 15: 5609–5626.

8. Bravo C, Moshou D, West J, McCartney A, Ramon H. Early Disease Detection in Wheat Fields using Spectral Reflectance. Biosystems Engineering. 2003; 84(2): 137–145.

9. Griffel L M, Delparte D, Edwards J. Using Support Vector Machines classification to differentiate spectral signatures of potato plants infected with Potato Virus Y. Computers and Electronics in Agriculture. 2018; 153: 318–324.

10. Jin X, Li Z, Feng H, Ren Z, Li S. Estimation of maize yield by assimilating biomass and canopy cover derived from hyper-spectral data into the AquaCrop model. Agricultural Water Management. 2020; 227: 105,846.

11. Li Z, Taylor J, Yang H, Casa R, Jin X, Li Z *et al.*, A hierarchical interannual wheat yield and grain protein prediction model using spectral vegetative indices and meteorological data. Field Crops Research. 2020; 248: 107,711.

12. Maimaitijiang M, Sagan V, Sidike P, Hartling S, Esposito F, Fritschi FB. Soybean yield prediction from UAV using multimodal data fusion and deep learning. Remote Sensing of Environment. 2020; 237: 111,599.

13. Yang Q, Shi L, Han J, Zha Y, Zhu P. Deep convolutional neural networks for rice grain yield estimation at the ripening stage using UAV-based remotely sensed images. Field Crops Research. 2019; 235; 142–153.

Chapter 16

Role of Bioinformatics in Crop Improvement

Sree Vathsa Sagar U.S.[1], Revanna Swamy K.M.[1], Ellandula Anvesh[1] and Divya Koripalli[2]*

[1]*Ph.D. Scholar (Department of Genetics and Plant Breeding), TNAU, Coimbatore, Tamil Nadu*
[2]*Ph.D. Scholar (Department of Genetics and Plant Breeding), University of Agricultural Sciences, Raichur, Karnataka*
**e-mail: sreevathsasagar@gmail.com*

INTRODUCTION

Bioinformatics is the integration of mathematical and computing approaches that are used for complete understanding of biological pathway and processes. It is a interdisciplinary study of biological materials that includes, genes, mRNA, proteins, metabolites and many physiological indices for using informatics tools, such as various algorithms and statistical methods. Specifically, highly complex biological data can be processed using advanced computer tools. The implementation of bioinformatics tools reduces the cost of complex analyses, thus providing scope for unorthodox areas like sustainable agriculture Ma³yska *et al.*, The typical datasets produced by plant researchers contain morphological, physiological, molecular, and genetic information that will describe the entire plant life cycle. Bioinformatics tools the collected data and extract key indices and trends to quickly and accurately generate hypotheses and then offer solutions.

In agricultural, the utilization of bioinformatics can help with efficient crop improvement and the plant resistance against pathogens and insects Gomez-Casati *et al.* scientists are interested to breeding and modifying crop species to improve the yield and quality, as well as creating new cultivars with traits that benefit human welfare and health. Bioinformatics accelerates the generation and development of these new varieties. Indeed, genes associated with specific traits can be analysed on a computer virtually before being introgressed into a plant, and these results can be analysed to determine which can be introduced further into the plant for a precise phenotypic analysis.

Each biological piece that has been measured at each level by a high-throughput method is represented in a plane in a concept model with different layers ranging from genome to phenome, this model is termed 'omic space' (Toyoda and Wada 2004). Such all comprehensive models often will provide an good initial point for designing an experiments, creating hypotheses or conceptualizing based on the integrated knowledge found in the omic space of a particular organism. Furthermore, development of such omic resources and data sets of various species allows the comparison of omic properties among related as well as unrelated species, which seems to be an efficient way to find accurate evidence for conserved gene functions that might be evolutionarily conserved. Bioinformatics platforms will be most essential tools for accessing omics data sets for the efficient mining and integration of biologically significant knowledge.

Bioinformatics plays a significant role in data integration, analysis, and model prediction, as well as in managing the massive amounts of data resulting from new, high-throughput approaches (Ko *et al*). Plant growth can be predicted based on the available wealth of physiological and phenotypic data, by creation of a virtual plant that can accurately predict growth patterns and the consequences of interactions with diseases or pests Kim *et al.*, Bioinformatics has also many applications in the management of plant resistance to various stresses.

Much study has been done on the molecular mechanisms underpinning plant responses to abiotic stress, and when coupled with the bioinformatics' predictive power, they may provide new opportunities for the agricultural sector. The following are significant bioinformatics subdisciplines:

- ☆ the creation and usage of technologies that make it possible to manage and utilise different forms of information with efficiency.

- ☆ The creation of new mathematical formulas (algorithms) and statistics that make it simple to analyse correlations between the constituents of larger data sets. Examples include techniques for finding a gene within a lengthy sequence, foretelling the structure and/or function of proteins, and grouping protein sequences into families of related sequences.

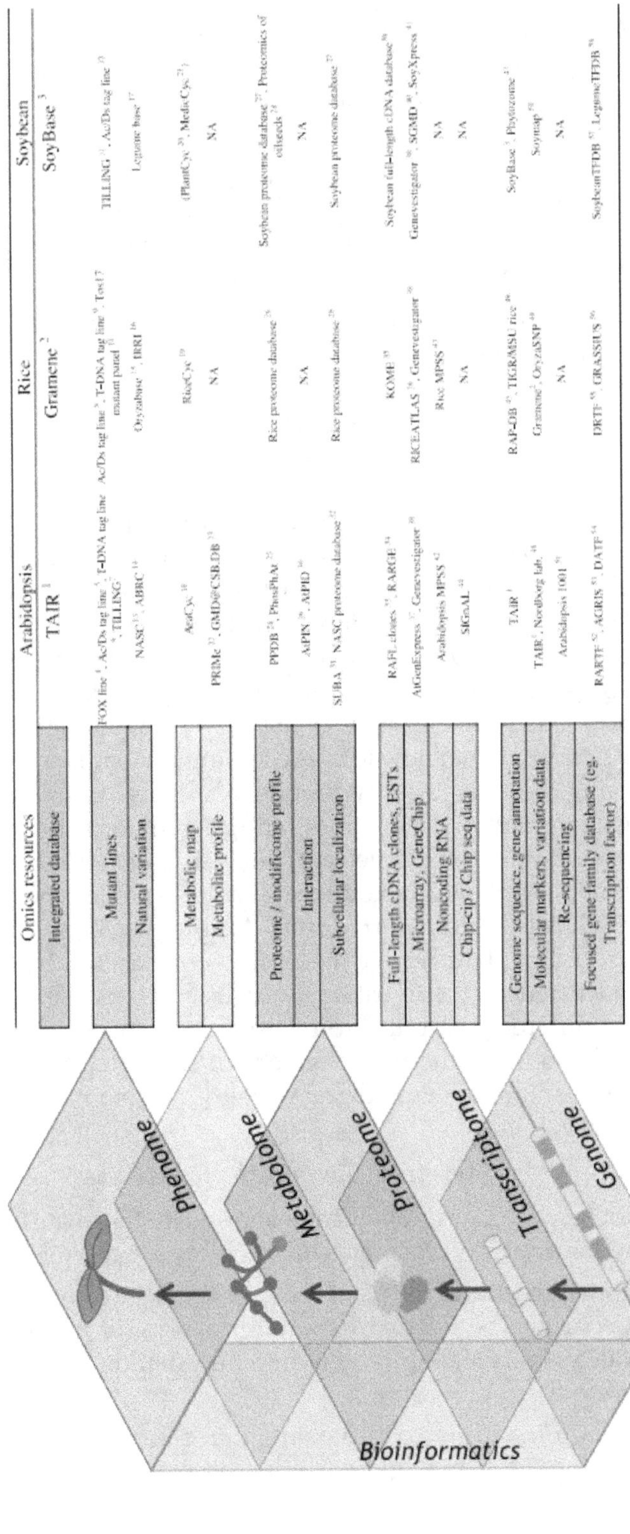

Omics resources	Arabidopsis	Rice	Soybean
Integrated database	TAIR [1]	Gramene [2]	SoyBase [3]
Mutant lines	FOX line [4], Ac/Ds tag line [5], T-DNA tag line [6], TILLING [7]	Ac/Ds tag line [8], T-DNA tag line [9], Tos17 mutant panel [10]	TILLING [11], Ac/Ds tag line [12]
Natural variation	NASC [13], ABRC [14]	Oryzabase [15], IRRI [16]	Legume base [17]
Metabolic map	AraCyc [18]	RiceCyc [19]	(PlantCyc [20], MedicCyc [21])
Metabolite profile	PRIMe [22], GMD@CSB.DB [23]	NA	NA
Proteome / modificome profile	PPDB [24], PhosPhAt [25]	Rice proteome database [26]	Soybean proteome database [27], Proteomics of oilseeds [28]
Interaction	AtPIN [29], AtPID [30]	NA	NA
Subcellular localization	SUBA [31], NASC proteome database [32]	Rice proteome database [28]	Soybean proteome database [27]
Full-length cDNA clones, ESTs	RAFL clones [33], RARGE [34]	KOME [35]	Soybean full-length cDNA database [36]
Microarray, GeneChip	AtGenExpress [37], Genevestigator [38]	RICEATLAS [38], Genevestigator [38]	Genevestigator [38], SGMD [40], SoyXpress [41]
Noncoding RNA	Arabidopsis MPSS [42]	Rice MPSS [43]	NA
Chip-cip / Chip seq data	SIGnAL [44]	NA	NA
Genome sequence, gene annotation	TAIR [1]	RAP-DB [45], TIGR/MSU rice [46]	SoyBase [3], Phytozome [47]
Molecular markers, variation data	TAIR [1], Nordborg lab. [48]	Gramene [2], OryzaSNP [49]	Soymap [50]
Re-sequencing	Arabidopsis 1001 [51]	NA	NA
Focused gene family database (eg. Transcription factor)	RARTF [52], AGRIS [53], DATF [54]	DRTF [55], GRASSIUS [56]	SoybeanTFDB [57], LegumeTFDB [58]

Bioinformatics — Phenome — Metabolome — Proteome — Transcriptome — Genome

Figure 16.1: Different Omic Represented in a Single Diagram from Phenomics to Genomics (Mochida *et al.*).

Importance of Bioinformatics

Bioinformatics' main objective is to improve our understanding of biological processes. To accomplish this, computationally costly procedures must be created and used. Sequence alignment, gene discovery, genome annotation, drug design, drug discovery, protein structure alignment, protein structure prediction, gene expression prediction, protein-protein interaction prediction, genome-wide association studies, and evolution modeling are the main informational and practical applications it serves.

Right now, bioinformatics involves the development of several databases, algorithms, computational and statistical approaches, and theoretical frameworks to address both theoretical and practical issues related to the management and interpretation of biological data.

Bioinformatics in Agriculture

From thousands of years, people are using breeding and selection to make domestic varieties of these crops with the wanted characteristics. Significant progress has been completed in taste, nutritional value and productivity, especially during the "Green Revolution" which took place in 1960 - 1970. Involvement of computer science in the area of plant biology has change the way we usually do research related to plants in previous decades. Rapid ground breaking progress of sequencing technology during the few last years made this technology so cost-effective that nowadays it is common for any experimental lab to use sequencing methods to study genome of interest. Within the field of plant improvement, agricultural genomics, or agri-genomics, bioinformatics is playing a significant role in the collection, analysis and storage of genomic data. Different ways through which bioinformatics tools and methods are used in agriculture, which is together referred to as Agri-informatics, primary aim is to improve the resistant plant against the biotic as well as the abiotic stresses and also for enhancement of the nutritional quality in crops. Gene discovery through the use of different computer software has also boosted scientists to develop specific methods for the improvement of seed quality, incorporate micronutrients into plants for enhanced grain quality, and engineer plants with phytoremediation capabilities.

There are many applications from bio informatics such as:

☆ Collection and storage of plant genetic resource those can be used for breeding for improved resistance of various pest and diseases.

☆ Comparative genetics of the plant species, where the completely known plant species can be utilized for understanding of unknow species which makes transferring of genomic information between related species very easy.

☆ Plant genomics aims to illustrate the genetic and molecular underpinnings of all biological activities in plants that are specific to each species. This knowledge is essential because it will enable the effective use of plants as biological resources to create new cultivars with higher quality and lower financial and environmental expenses. Pathogen and abiotic stress tolerance, plant quality features, and reproductive qualities affecting yield are the traits of major importance.

Table 16.1: Whole Genome Sequencing Projects in Plants

Common Name	Latin Name	Status	Sequencing Group	Sequencing Method
Mouse ear cress	*Arabidopsis thaliana*	Completed	Consortium (AGI)	clone
Poplar	*Populus trichocarpa*	Completed	JGI	whole genome shotgun
Lyreleaf rockcress	*Arabidopsis lyrata*	In progress	JGI	whole genome shotgun
Pink shepherd's purse	*Capsella rubella*	In progress	JGI	whole genome shotgun
Rapeseed	*Brassica rapa*	In progress	Consortium (MGBP)	clone
Tomato	*Lycopersicum esculentum*	In progress	Consortium (ITGSP)	clone
Potato	*Solanum tuberosum*	In progress	Consortium (PGSC)	clone
Barrel medic	*Medicago truncatula*	Completed	Consortium (IMGAG)	clone
Lotus	*Lotus japonicus*	In progress	Consortium	clone
Soybean	*Glycine max*	In progress	JGI	whole genome shotgun
Cotton	*Gossypium hirsutum*	In progress	JGI	whole genome shotgun
Cassava	*Manihot esculenta*	In progress	JGI	whole genome shotgun
Grape	*Vitis vinifera*	Completed	Consortium	whole genome shotgun
Columbine	*Aquilegia formosa*	Pending	JGI	whole genome shotgun
Eucalyptus	*Eucalyptus grandis*	Pending	JGI	whole genome shotgun
Papaya	*Carica papaya*	In progress	Consortium	whole genome shotgun
Castor bean	*Ricinus communis*	Completed	TIGR	whole genome shotgun
Yellow owl's clover	*Triphysaria versicolor*	Pending	JGI	whole genome shotgun

Common Name	Latin Name	Status	Sequencing Group	Sequencing Method
Monocot				
Rice	*Oryza sativa japonica Nipponbare*	Completed	Consortium (IRGSP)	clone shotgun
Rice	*Oryza sativa indica*	Completed	Beijing Genomics Institute	whole genome shotgun
Maize	*Zea mays*	In progress	Consortium	clone
Sorghum	*Sorghum bicolor*	Completed	JGI	whole genome shotgun
Grass	*Brachypodium distachyon*	In progress	JGI	whole genome shotgun
Foxtail millet	*Setaria italica*	Pending	JGI	whole genome shotgun
Banana	*Musa acuminata*	In progress	Consortium	clone
Wheat	*Triticum aestivum*	Pilot	Consortium (IWGSC)	clone or single chromosome shotgun

The whole genome sequencing data of different plant species have a played a very significant role improvement of many crop species starting it with the model crop like Mouse ear cress (*Arabidopsis thaliana*), where it acted as the basis for other genome sequencing, gradually many plant species were sequenced that include both commercial viable plants and others as well. This data will be source for many important traits like quality traits, insect resistance, disease resistance and many other important traits. This can be used for trait introgression from wild species or the related species to desired genetic background.

☆ Additionally, the development of new plant diagnostic and analysis technologies depends on this knowledge. Pathogen and abiotic stress tolerance, plant quality features, and reproductive traits defining production are traits that are deemed to be of the utmost importance. Currently, a genome program can be seen as a crucial instrument for improving crops.

Bioinformatics Tools mostly used

Homology and similarity tools, protein functional analysis tools, sequence analysis tools, and other tools are several categories of bioinformatics tools. Both visualization tools and data-mining software are available to analyze and retrieve information from proteomic databases, which store genetic sequence data.

Some examples of Bioinformatics Tools:

1. BLAST (Basic Local Alignment Search Tool)
2. Fasta (FAST homology search All sequences)
3. EMBOSS (European Molecular Biology Open Software Suite)
4. Clustalw
5. RasMol
6. PROSPECT (PROtein Structure Prediction and Evaluation Computer ToolKit)
7. Pattern Hunter
8. COPIA (COnsensus Pattern Identification and Analysis)

Table 16.2: Bioinformatics Tools and Websites that can be used in Plant Research

Database Name	URL	Note
Tbtools	https://github.com/srbehera11/stag-cns	An integrated toolkit for interactive analysis of big biological data
SMART	http://smart.embl-heidelberg.de/	Protein conserved domain prediction tool
STAG-CNS	https://github.com/srbehera11/stag-cns.	A sequentially conserved non-coding sequence discovery tool for an arbitrary number of species
FED	http://www.hi-tom.net/FED.	Genome editing exogenous component detection platform
MAFFT	https://mafft.cbrc.jp/alignment/server/	Online sequence matching tool
Protter	http://wlab.ethz.ch/protter/start/	Online protein structure mapping tool
EvolView	https://www.evolgenius.info/evolview/	Web-based tools for visualizing, annotating, and managing system trees
iTOL	https://itol.embl.de/	Online tool for displaying, managing, and annotating system development trees

Many bioinformatic tools available for plant research Tbtools where it will be used for visualization of the data along with data editable features. Protter database is extensively used for the protein structure visualization and prediction of the proteomic data.

Table 16.3: Bioinformatics Tools Specific to Applications in Plant Research

Database Name	URL	NOTE
BAR	**http://www.bar.utoronto.ca/welcome.htm** Accessed on 4 May 2022.	Plant biology analysis tools platform
CRISPR-P	**http://crispr.hzau.edu.cn/CRISPR2/** Accessed on 4 May 2022.	Improved CRISPR/Cas9 toolkit for plant genome editing
ACT	**https://www.michalopoulos.net/act/** Accessed on 4 May 2022.	Arabidopsis co-expression analysis tool
OryGenesDB	**http://orygenesdb.cirad.fr/** Accessed on 4 May 2022.	An interactive tool for reverse genetics studies in rice
T-DNA Express	**http://signal.salk.edu/cgi-bin/tdnaexpress** Accessed on 4 May 2022.	Arabidopsis gene targeting tool
Plant MetGenMAP	**http://bioinfo.bti.cornell.edu/cgi-bin/MetGenMAP/home.cgi** Accessed on 4 May 2022.	Web-based tools for comprehensive mining and integration of gene expression and metabolite changes in the context of biochemical pathways
iTAK	**http://itak.feilab.net/cgi-bin/itak/index.cgi** Accessed on 4 May 2022.	Software packages for identifying and classifying plant transcription factors and protein kinases
PlantPAN	**http://plantpan2.itps.ncku.edu.tw/** Accessed on 4 May 2022.	Tools for detecting transcription factor binding sites in plants
SnpHub	**http://guoweilong.github.io/SnpHub/** Accessed on 4 May 2022.	A unified web server framework for exploring large-scale genomic variation data

Many databases specific for the plant research are extensively used in plant research like PlantPAN where it is utilized for finding the transcription factor binding sites. TRANSFAC and PlnTFDB are the database for the transcription factors in the plants. These will predict the TF binding sites in the promoter region of the plants.

Application of Bioinformatics in Plant Breeding

Plant breeders always work towards the improving cultivars which will eventually solve the major problems of the agriculture through developing the new cultivars, thus for that it will start with basics research and information like taxonomy to genomic data, where the genomic assisted breeding is one of the effective methods for precise crop improvement and cost effective than conventional, genomic information will provide the overall idea about the organisation and potential of the genotypes. Even the genomic information form the different related species can also be used for improvement. These tools can be employed in many ways like:

1. Breeding for High Yield and Quality

Main and primary aim of any breeder would be to increase the yield and quality, where that can achieved easily using these tools, by using the bioinformatic analysis that are related to the yield and quality can be screeded and obtained which can be taken from either related species or the closely related ones. Even with these tools we can even have ideal plant traits which can yield more and efficient in plants, for traits like leaf angle, number of leaves, plant architecture can be designed before selection using the bioinformatics.

2. For Prediction of Plant Growth and Conditions

Improvement of traits like yield depends on many other traits as it being the polygenic traits, so it depends on many traits, thus understanding these is very important. By using the bioinformatics tools we can predict the yield and other traits for example leaf angle is very important trait that can effect the photosynthetic efficiency, so basically this will in turn affect the spacing in plant, so to have less spacing and more plants, and more yield we need to have leaf angle which can be visualized using the bioinformatic tools and this is true even for the many abiotic stress reacted traits, thus we can early standardize and visualize the traits before going for actual selection.

3. Automation for Agriculture

Currently automation have taken over all the sectors like this can be used in the agriculture as well where we can reduce the errors and increase the efficiency of agriculture practices for example where we can unnecessary nutrient miss managements of plants, as electric sprayer can spray the at each hill precisely, that according to the requirement of the plant so no extra application will be wasted. For this bioinformatics tools can be used, bioinformatic breeder cultivars are cultivars where the traits which needs to transferred to donor parent will be processed before transferring to the recipient parent. This new breeding method operation is simple, low cost, and does not destroy the organism's own genes, this bioinformation breeder is based on a special kind of bioenergy– biomicrowave

4. Precise Prediction of Experimental Results and Transgenic Phenotypes

At present in plant research the genotype and phenotype prediction have been conventionally done using the statical methods by statistical methods, two methods like autoregressive (AR) and Markov chain (MCMC), are used to predict the growth trends of plants by using the Normalized Difference Vegetation Index (NDVI). But application of machine learning to prediction of phenotype-genotype will facilitate the study of the roles played by various molecular components in shaping plant phenotypes, breeders mostly depend on the genomic data for their selection and QTL mapping is prerequisite for that, many researchers have used the machine learning tools to map QTL, although the application is very less as of now nut it will be more in coming days and increase its use.

Databases that are Used for Crop Improvement

Genome Sequencing Projects

The first genome sequence of a plant was done for *Arabidopsis thaliana*, which is now considered as a model species in plant biology due to its small size, short generation time and high efficiency of transformation. The genome sequencing was published in 2000 by the Arabidopsis Genome Initiative (AGI) (The Arabidopsis Genome Initiative 2000). The draft genome sequence of rice, both japonica and indica, an important food crop as well as a model monocotyledon, was published in 2002 (Goff *et al.*, 2002, Yu *et al.*, 2002). Subsequently, the genome sequence of japonica rice was completed and published by the International Rice Genome Sequencing Project in 2005 (International Rice Genome Sequencing Project 2005). To date, several genome sequencing projects involving various plant species have been completed (Table 16.1).

Some of the most widely used plant bioinformatics online databases are Gateway of Brassica Genome, BGI Rice Information System, ChloroplastDB, The Crop Expressed Sequence Tag (CR-EST_ database, CyanBase, the European Molecular Biology Laboratory (EMBL) nucleotide sequence database, and many more.

Plant Comparative Genomics and Databases

The recent discovery of nucleotide sequences for plant species, including crops, now we can perform genome-wide comparative analyses of model organisms with the aim of discovering key genes involved in phenotypic characteristics (Neale and Ingvarsson 2008). The integration of genomic database derived from various related species, such as large-scale collections of cDNAs and data from whole-genome sequencing projects, will facilitate sharing of information about gene function between models and applied organisms represented in Table 16.4.

Table 16.4: Database List and URL Link for the Databases

Database Name	Species	URL
TAIR	Arabidopsis	http://www.arabidopsis.org/
SIGnAL	Arabidopsis	http://signal.salk.edu/
RARGE	Arabidopsis	http://rarge.psc.riken.jp/
Rice Genome Annotation Project	Rice	http://rice.plantbiology.msu.edu/
RAP-DB	Rice	http://rapdb.dna.affrc.go.jp/
SOL genomics network	Solanaceae	http://solgenomics.net/
Gramene	Gramineae	http://www.gramene.org/
GrainGenes	Triticeae and Avena	http://wheat.pw.usda.gov/GG2/index.shtml
SoyBase	Soybean	http://www.soybase.org/
MazieGDB	Maize	http://www.maizegdb.org/
CyanoBase	Cyanobacteria	http://genome.kazusa.or.jp/cyanobase/
GDR (Genome Database for Rosaceae)	Rosaceae	http://www.bioinfo.wsu.edu/gdr/
Brassica Genome Gateway	Brassica	http://brassica.bbsrc.ac.uk/
Cucurbit Genomics Database	Cucurbitaceae	http://www.icugi.org/
Phytozome	whole species genome	http://www.phytozome.net/
PlantGDB	whole species genome and/or large-scale EST data available	http://www.plantgdb.org/
EnsemblPlants	Plant species (whole genome data available)	http://plants.ensembl.org/index.html
ChloroplastDB	Plant species (Chloroplast genome data available)	http://chloroplast.cbio.psu.edu/
KEGG PLANT	Plant species (whole genome and/or large-scale EST data available)	http://www.genome.jp/kegg/plant/

☆ TAIR is a well-known and comprehensive information source for plant science and a crucial database for Arabidopsis research.

☆ The Salk Institute Genomic Analysis Laboratory (SIGnAL) is a resource for information that compiles a number of data sets of important omics findings pertaining to Arabidopsis. The RIKEN Arabidopsis Genome Encyclopedia (RARGE) provides information on various genomic resources built at RIKEN for Arabidopsis research

☆ A well-known portal site called Gramene promotes plant comparative genomics in addition to serving as an integrated resource for information about rice.

✩ According to Liang *et al.* (2008), Gramene provides integrated genome-oriented data, such as gene annotation and molecular markers, as well as a QTL database that is primarily for Gramineae species.

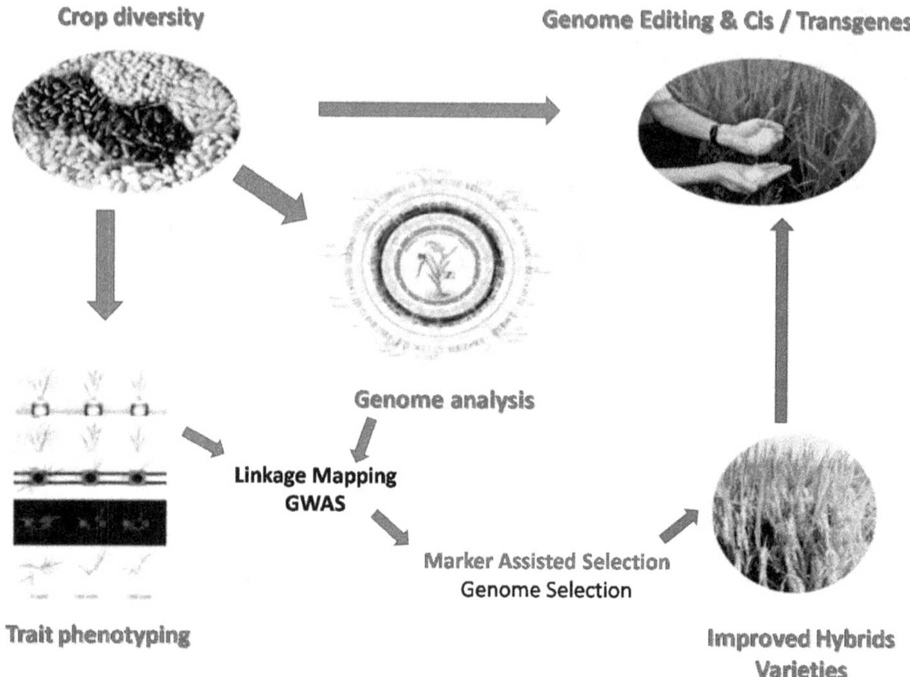

Figure 16.2: Flow Chart of different Events that are Involved in Crop Improvement

Crop improvement majorly relays on genotypic as well as phenotypic data for accuracy, thus markers developed will be used for the selection of desired genotypes indirectly, there are a large set of genotypic data obtained form across all the genotypes of species, as this will be a very huge data, management of this data will be tedious and important. Form the fig. 1 we can observe that equal importance must be given to phenotyping as well as genotyping, this will require advanced data management strategies that involves many bioinformatic tools, which will make use of that to integrate both the data to give output after analysis, that can be very helpful in obtaining the insight of the genotypes which we are analysing and selection of genotypes based on objective.

Conclusion

Bioinformatics is a method to understand and analyse the biological data using the mathematical and computing tools. Here, in this chapter it is all about the application different bioinformatic tools for agriculture and specially in crop improvement, bioinformatics plays a big role in providing the tools for integrating different data sets, which are completely different where they can't be interpreted normally, that's where the use of the these advanced tools come into play. Although generating the database is the crucial step in this, different data at different level of organisms organization like at genomic level (genomics), RNA level (transcriptomics), metabolic level (metabolomics), epigenetic study (epigenomics) and others, So these are completely different from each other in both data format and the information they give, thus to integrate and interpret we need these tools, which make our interpretation and prediction about either pathway or genotype easy and increased accuracy. As the number of genotypes increases then the data size also increases that intern makes analysis of these data difficult, so by using the mathematical algorithms and software, it will make it simple. There are also many different databases for different crops, many of them are online databases that are always available, thus use of bioinformatics and it tools will have very fruit full effects in crop improvement.

REFERENCES

(http://www.arabidopsis.org/portals/genAnnotation/other_genomes/index.jsp)

Aditya Shastry, K.; Sanjay, H.A. Hybrid prediction strategy to predict agricultural information. Appl. Soft Comput. 2020, 98, 106811.

Goff SA, Ricke D, Lan TH, Presting G, Wang R, Dunn M, *et al.*, A draft sequence of the rice genome (Oryza sativa L. ssp. japonica), Science, 2002, vol. 296 (pg. 92-100)

Gomez-Casati, D.F.; Busi, M.V.; Barchiesi, J.; Peralta, D.A.; Hedin, N.; Bhadauria, V. Applications of Bioinformatics to Plant Biotechnology. Curr. Issues Mol. Biol. 2018, 27, 89–104.

Iida K, Seki M, Sakurai T, Satou M, Akiyama K, Toyoda T, *et al.*, Genome-wide analysis of alternative pre-mRNA splicing in Arabidopsis thaliana based on full-length cDNA sequences, Nucleic Acids Res., 2004, vol. 32 (pg. 5096-5103)

International Rice Genome Sequencing Project The map-based sequence of the rice genome, Nature, 2005, vol. 436 (pg. 793-800)

Itoh T, Tanaka T, Barrero RA, Yamasaki C, Fujii Y, Hilton PB, *et al.*, Curated genome annotation of Oryza sativa ssp. japonica and comparative genome analysis with Arabidopsis thaliana, Genome Res., 2007, vol. 17 (pg. 175-183)

Keiichi Mochida, Kazuo Shinozaki, Genomics and Bioinformatics Resources for Crop Improvement, Plant and Cell Physiology, Volume 51, Issue 4, April 2010, Pages 497–523.

Kim, T.; Lee, S.-H.; Kim, J.-O. A Novel Shape Based Plant Growth Prediction Algorithm Using Deep Learning and Spatial Transformation. IEEE Access 2022, 10, 37731–37742.

Ko, G.; Kim, P.-G.; Yoon, J.; Han, G.; Park, S.-J.; Song, W.; Lee, B. Closha: Bioinformatics workflow system for the analysis of massive sequencing data. BMC Bioinform. 2018, 19, 43.

Liang, C., Jaiswal, P., Hebbard, C., Avraham, S., Buckler, E.S., Casstevens, T., *et al.* (2008) Gramene: a growing plant comparative genomics resource. Nucleic Acids Res. 36 : D947 – D953.

Ma³yska, A.; Jacobi, J. Plant breeding as the cornerstone of a sustainable bioeconomy. New Biotechnol. 2018, 40, 129–132

Nanjo T, Sakurai T, Totoki Y, Toyoda A, Nishiguchi M, Kado D, *et al.*, Functional annotation of 19,841 Populus nigra full-length enriched cDNA clones, BMC Genomics, 2007, vol. 8 pg. 448

Neale DB, Ingvarsson PK. Population, quantitative and comparative genomics of adaptation in forest trees, Curr. Opin. Plant Biol., 2008, vol. 11 (pg. 149-1

Ralph SG, Chun HJ, Cooper D, Kirkpatrick R, Kolosova N, Gunter L, *et al.*, Analysis of 4,664 high-quality sequence-finished poplar full-length cDNA clones and their utility for the discovery of genes responding to insect feeding, BMC Genomics, 2008, vol. 9 pg. 57

The Arabidopsis Genome Initiative Analysis of the genome sequence of the flowering plant Arabidopsis thaliana, Nature, 2000, vol. 408 (pg. 796-815)

Umezawa T, Sakurai T, Totoki Y, Toyoda A, Seki M, Ishiwata A, *et al.*, Sequencing and analysis of approximately 40 000 soybean cDNA clones from a full-length-enriched cDNA library, DNA Res., 2008, vol. 15 (pg. 333-346)

Yamamoto YY, Yoshitsugu T, Sakurai T, Seki M, Shinozaki K, Obokata J. Heterogeneity of Arabidopsis core promoters revealed by high-density TSS analysis, Plant J., 2009, vol. 60 (pg. 350-362.

Yu J, Hu S, Wang J, Wong GK, Li S, Liu B, *et al.*, A draft sequence of the rice genome (Oryza sativa L. ssp. indica), Science, 2002, vol. 296 (pg. 79-92).

Index